WACO-McLENNAN COUNTY LIBRARY
1717 AUSTIN AVE
WACO TX 76701

underground cities

First published in 2020 by Frances Lincoln Publishing,
an imprint of The Quarto Group.
The Old Brewery, 6 Blundell Street
London, N7 9BH,
United Kingdom
T (0)20 7700 6700
www.QuartoKnows.com

Text © 2020 Mark Ovenden
Illustrations © 2020 Robert Brandt
Map Illustrations by Lovell Johns

Mark Ovenden has asserted his moral right to be identified as the Author of this Work in accordance with the Copyright Designs and Patents Act 1988.

All rights reserved. No part of this book may be reproduced or utilised in any form or by any means, electronic or mechanical, including photocopying, recording or by any information storage and retrieval system, without permission in writing from Frances Lincoln Publishing.

Every effort has been made to trace the copyright holders of material quoted in this book. If application is made in writing to the publisher, any omissions will be included in future editions.

A catalogue record for this book is available from the British Library.

Publisher: Richard Green
Art Director: Paileen Currie
Layout: Mark Ovenden
Editorial Director: Jennifer Barr
Commissioning Editor: Lucy Warburton
Senior Editor: Laura Bulbeck
Copy Editor: Anna Southgate
Indexer: Helen Snaith

ISBN 978-1-78131-893-5
Ebook ISBN 978-1-78131-894-2

10 9 8 7 6 5 4 3 2 1

Printed in China

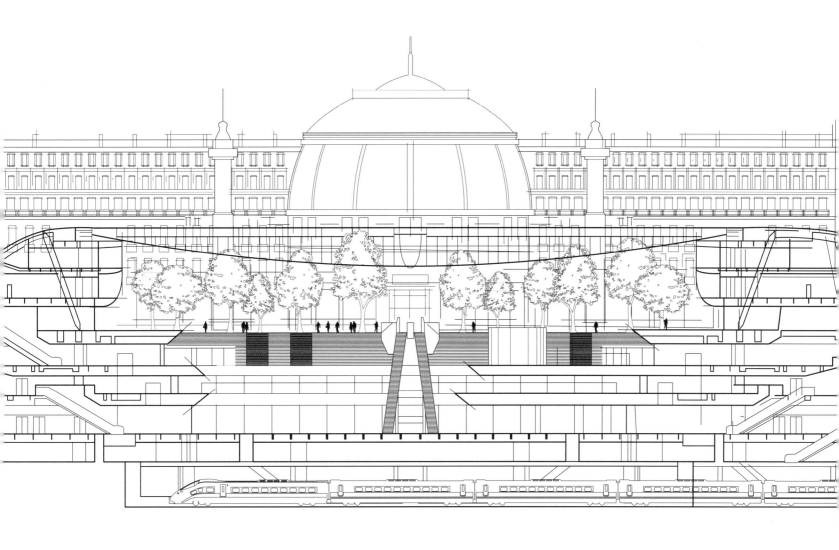

underground cities

mapping the tunnels, transits and networks underneath our feet

MARK OVENDEN

Starting at the International Date Line, take a trail around the world to discover how these 32 cities use their underground spaces.

Contents

Introduction 6

NORTH AND SOUTH AMERICA

Los Angeles.................. 10
Made for streetcars

Mexico City.................. 18
The city in the dried up lake

Chicago.................. 26
The elevated city

Cincinnati.................. 36
The subway that never happened

Toronto.................. 38
Weatherproof shopping

Montreal.................. 46
Bilingual basements

New York City.................. 54
The global capital

Boston.................. 64
Tea Party was just the start

Buenos Aires.................. 72
Persecution and perseverance

EUROPE

Gibraltar.................. 80
The riddled rock

Madrid.................. 82
Mazes and metros

Liverpool.................. 86
First rail tunnels

Manchester.................. 90
Pioneers and pipedreams

London.................. 94
On Roman shoulders

Barcelona..................104
The planned city

Paris..................112
The Swiss cheese of Europe

Rotterdam..................122
Holding back the sea

Amsterdam..................126
Hidden under canals

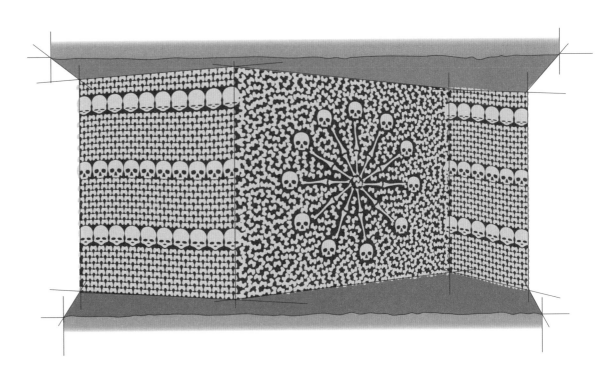

Marseille	132
Tunnels made a beach	
Milan	134
Crypts and pieces	
Oslo	138
Opportune opening	
Rome	142
Where roofs become foundations	
Munich	150
From the ashes	
Berlin	152
Divided and healed	
Budapest	160
Thermal layers	
Stockholm	168
In love with tunnels	
Helsinki	176
Sheltering an entire city	
Moscow	182
Secret subterrania	

ASIA AND OCEANIA

Mumbai	194
City of seven islands	
Beijing	196
Created by hand	
Tokyo	200
Meeting under a megalopolis	
Sydney	208
Roads to nowhere	

Introduction

Walking around most cities in the twenty-first century, people's feet are highly likely to be falling on top of a lot more than the pavement. If it were possible to neatly peel away the soil, among the worms and moles would be found a messy, yet mesmerizing mix of cable ducts, utility pipes, drainage channels, cellars, crypts, wells, tunnels, subways and foundations of older buildings. In the longer established cities, we are quite literally standing upon the shoulders of giants of civilizations past.

Why do societies continually build new structures on top of old ones and what happens to their forgotten remains? How much of an original Roman or prehistoric settlement lies buried beneath a modern city? What will be found once a new subway line is bored through the earth? Questions like these have fascinated me since I was a kid being dragged somewhat unwillingly around the detritus of de-industrializing Britain in the 1960s and 70s by my studious father who delighted in long walks along abandoned railways or disused canals.

Shortly after relocating from the city where I was brought up to a rural village, the pupils at my new school were asked to compose a quatrain about their home town. I wrote:

I come again to London Town
Green land it has diminished
New buildings up, old landmarks down
When will London be finished?

Not the makings of a future poet, but the beginnings of recognition that constant renewal of cities seems to have placed in my psyche a drive to learn more about how urban areas are on a never-ending cycle of regeneration. The construction of public transportation systems and new building foundations are arguably the leading causes of unearthing buried human activity, so naturally being a transit geek led me to question more about what else lies down there beyond the subway walls.

For most of my life I collected transport ephemera (maps, brochures, photos, etc.) but during the 1990s I was lucky enough to take more international trips for my work in the music industry, so was able to add worldwide examples to my collection. Friends going to far flung cities began asking for copies of the Mexico City or Tokyo Metro map and I got to wonder why there was no book which showed all of them side by side. This led me to develop ideas for my first publication *Metro Maps of the World* in 2003. It was a bit of a surprise hit and was taken on by bigger publishers, translated into other languages and updated several times since. The success also inspired a move to France to investigate the Paris Métro, and another back to London in time to produce a book on the design of the Underground for its 150th anniversary in 2013. All of this interest and new knowledge led to more intensive urban exploration and the premise of *Underground Cities* which examines all types of sub-surface human-made structures, not just my traditional fare of transit tunnels.

The research for this has led to some surprising discoveries. For example, the vast complex of spaces hewn out beneath Helsinki, begun during the Cold War as a way of protecting the citizens from attack, has been continually maintained and expanded to the point that it now functions as another working level of this fine city. On a more commercial note are the vast interlinked pedestrian tunnel networks of the Canadian shopping malls beneath cities like Montreal and Toronto. And although I was aware of Moscow's rumoured second, deep-level, military Metro, I had no idea about the scale of some of the secret bunkers created there. These and many other findings are assembled in this book.

The richness and diversity of those sites has been embellished by superb illustrations drawn by Robert Brandt. His interpretations of these underground places are made more comprehensible by the clarity of his work. Thanks must also go to Lovell Johns, in particular to Clare Varney who created the wonderful maps.

The order in which to display the cities presented some challenges, but rather than showing them alphabetically or by population size, we settled on circumnavigating the globe from the International Date Line to avoid any accusation of favouritism!

From subsurface swimming pools to clandestine control rooms, the diversity of features shown in this book should inspire anyone to wonder about the labyrinths underfoot. Whether this is being read from the perspective of someone keen to travel to these cities, or just as an armchair explorer, I hope that you will experience the fascination of what lies beneath our feet as much as I have and next time you walk along a city street, your imagination of the invisible is expanded.

North and South America

A 1906 postcard showing a cutaway of the BMT Subway and
Brooklyn Bridge Terminal in New York City

Los Angeles

Made for streetcars

One of the world's most extensive urban regions, with a population of four million people (and more than twelve million in the metropolitan area), Los Angeles lies on the Pacific Coast of California, USA. Infamous for its freeways and cars, the city was, surprisingly, laid out with mass transportation in mind.

Modern Los Angeles, commonly shortened to LA, sprawls out over a mainly flat plain in front of the Santa Monica and San Gabriel mountain ranges. Native peoples lived in the area from around 5000 to 3000 BCE, and Spanish explorers first arrived in 1542, returning 200 years later, when it is thought about 5,000 Gabrielinos lived in the area that is now known as the Los Angeles basin. The Town of Our Lady the Queen of the Angels was founded by forty-four settlers in 1781, although it was little more than a small ranch. By 1820 it had a population of 650 and fell to Mexican rule, but this ended in 1847. Railways arrived when the Southern Pacific was completed in 1876, followed by the Sante Fe in 1885. Oil was discovered in 1892 and by 1900 the population had topped 100,000.

LA's semi-arid position

Although not as dry as popular culture would have people believe, by the end of the nineteenth century, the Los Angeles basin was not receiving enough rain to provide drinking water to sustain its population. The Los Angeles Aqueduct project was intended to drain water from the Owen Valley north of the city, via 60km (37 miles) of canals, 70km (43 miles) of underground culverts and 156km (97 miles) of open, lined channels. With work having begun in 1905, the immense task was completed in 1913. It was so effective that nearby towns and cities such as Hollywood, also suffering from its own water shortages, unilaterally annexed themselves to LA to benefit from the works. This in turn caused more water usage than was anticipated and the project had to be extended in later years – for example, by the Mono Basin extension in 1930 and the Second Aqueduct in 1956 – so that it now draws water from almost as far as the state border with Nevada.

The 82-km (51-mile) Los Angeles River drains snowmelt and rain from the mountain ranges behind the city. Frequent flooding of the plain on which LA expanded led the city to engage the US Army Corps in straightening the river during the mid-1930s, directing it into a wide concrete channel. The stark infrastructure – a 'water freeway' – has been the backdrop of countless films and TV shows from *Grease* to *Terminator 2*. Today, the city is working with architects such as Frank Gehry to replace much of the concrete with natural grassed terraces in order to introduce a more environmental look in the run up to the city hosting the 2028 Summer Olympic Games.

Streetcars and subways

Although the first network of streets on the Santa Clara plain arose from Spanish planning doctrines, and was therefore laid out in a grid-like pattern

ABOVE LEFT Many miles from the metropolis, the Los Angeles Aqueduct collects water and funnels it into the city.

ABOVE RIGHT Part of the Pacific Electric tunnels which led to the Subway Terminal Building, opened in 1925. It carried an average of 65,000 passengers every day. This photo was taken around 1945, just a decade before it was abandoned.

with an open central plaza, the continuation of this style became a huge boon to the first public transportation in LA – the horse-drawn bus – which began in 1873. The Pioneer Omnibus Street Line was, however, dogged by rutted road surfaces, so the next year rails were laid along Spring and Sixth streets. The ease of crossing the city via streetcar gave rise to property speculation at the fringes and, in 1876, East LA (now Lincoln Heights) and Boyle Heights became the first suburbs as the new streetcar tracks arrived, still pulled by horses. Another company, the Los Angeles and Aliso Avenue Street Passenger Railway also served Boyle Heights in 1877, and this pattern of suburbs, established around streetcar lines, was widely emulated, utterly transforming the once wild and wooded plain to the modern city layout that exists today.

By the 1880s, new technology made it possible to power streetcars in other ways. In 1885, for example, an underground trench was dug for a cable to be buried and the Second Street Cable Railway was easily able to use it to pull vehicles up the steep inclines of Bunker Hill on which horses struggled. This opened up the hilly western area for new suburbs. Electric power arrived in 1887 and overhead wires appeared all over LA. The last horse-drawn streetcar was pulled just a decade later by the Main Street and Agricultural Park Street Railroad company.

The company operating narrow-gauge streetcars in the downtown area and nearby local neighbourhoods was the Los Angeles Railway (LARy, also known as Yellow Cars and, later, Los Angeles Transit Lines). They were established in 1901, when a real-estate tycoon called Henry E. Huntingdon bought a number of pre-existing companies. Merging their lines with his, he grew the network to an intensively used twenty-five lines, served by distinctive canary-yellow-painted vehicles.

Huntingdon realized that, although the Yellow Cars were perfectly suited for mass transit within the city centre, connecting the now sprawling suburbs of the whole basin required something more robust. In 1901, he formed the Pacific Electric Railway Company with the intention of consolidating the operations of smaller, outlying companies and laying new tracks between them, all at a standard gauge that would support a larger inter-urban tram. Their bright-crimson street-running trams soon became known as the Red Cars. An aggressive takeover policy culminated in 1911, with the Great Merger, creating 'new' Pacific Electric. It became the world's largest electric railway system during the 1920s with over 1,600km (994 miles) of track running Red Cars throughout Southern California. So was born the urban backbone of the state.

Pacific Electric was soon running so many services from outlying areas into downtown LA that the company considered it worthwhile to spend millions of dollars on the creation of the Subway Terminal Building that would be reached via long tunnels. When it opened in 1925, the Belmont Tunnel was effectively the first underground in the city, cutting around fifteen minutes off the time it once took to get from the downtown area to its far-flung outlying areas. Little of the building remains today, but it once had six platforms under Hill Street, from where the double-track tunnels took streetcars almost 2km (1.3 miles) west to the portal at Glendale Boulevard. The line quickly became known as the Hollywood Subway. Although the streetcars were shut down in the 1960s, the huge underground spaces found a second life as a nuclear fallout shelter during the Cold War.

The rise of the automobile

Possibly inspired by the building of the Hollywood Subway, the City and County of Los Angeles did some 'blue-sky-thinking' of its own. The Comprehensive Rapid Transit Plan of 1925, which envisaged creating a series of elevated lines similar to those in Boston and Chicago, would also have had them diving into new tunnelled sections in the downtown area and the immediate vicinity. In the suburbs, the plan was to take over some key Pacific Electric lines, and widened them onto 'grade-separated' sections to aid their flow away from automobile traffic. At least four of these lines were proposed for 'immediate construction', but no sod was turned. Although the plan was revisited and massively paired down after the end of the Second World War, the only surviving suggestion for underground sections was to repurpose the Hollywood Subway and add a short new section of tunnel under Broadway. This too came to nothing.

Even by the 1930s, traffic congestion between streetcars and privately owned motor vehicles was reaching crisis point. The Automobile Club of Southern California called for an elevated freeway system with the express aim of replacing the streetcars with buses. However, when the city began drawing up plans for freeways, a double-track space was set aside in the central reservation for rail-based services. The streetcars went into decline as car use soared, and by the early 1950s land grabs to build the biggest new freeways were in full swing. The great inter-urban network of the Red Cars and intense local Yellow Car services were mostly abandoned and the remaining streetcar routes were sold off to a company called National City Lines. It later emerged that this company was sponsored by General Motors directors with the specific intent to run down and abandon all streetcars so that people would be forced to buy more cars. The Great American Streetcar Scandal forms the premise of the 1988 film *Who Framed Roger Rabbit*.

The LA Metro

Despite the loss of its first section of subway, LA was not done with transit tunnels. A 1954 Los Angeles Metropolitan Transit Authority plan proposed a monorail line from Panorama City to Long Beach, with a subway section running below downtown Hill Street. By 1960, despite some preliminary test bores, nothing concrete had materialized and a new plan revived the elevated lines concept. Two years later, an east–west idea suggested a 'Backbone Route', which included a long tunnel to Century City beneath Wilshire Boulevard and an elevated section above the San Bernadino Freeway. As the 1960s dragged on with no stone turned, other ideas surfaced, including, in 1968, a 'five-corridor system', which looked remarkably similar to what eventually transpired with the city's metro network, although no work was carried out at the time.

The situation couldn't last, and in the mid-1970s, the city woke up to the desperate need for mass transit. Several proposals were discussed to build a full-scale subway for the car-addicted city now choking in exhaust fumes and obscured by regular smogs. Finally, in 1980, voters approved a local tax to support the building of the Los Angeles Metro Rail network.

After years of design work, construction began in 1985. In July 1990, the Metro's Blue Line was opened to the public. It had a short section in tunnel, where it was to provide interchange with the Red Line, which was completed three years later and was mostly in tunnel. The predominantly elevated Green Line opened in 1995. Several stations – for example Hollywood/Vine – are beautifully decorated underground cathedrals.

Bunker Hill

The focus of the 1970s LA Downtown People Mover was to create a 2.5-km (1.6-mile) rapid transit line with four stations, one underground (near 3rd and Olive Streets) in a tunnelled section that would have run between Flower and Hill Streets. Construction of the tunnel was already underway, when President Ronald Reagan's administration halted the project in 1981. A small section along 3rd Street survives. Had it gone ahead, the Bunker Hill Transit Tunnel may have gone on to connect to the LA Metro network, already in the pipeline.

In the same vicinity of Bunker Hill, several other tunnels remain open to the public. Allegedly constructed at various times during the 1900s, they mainly link a series of government buildings in the area. Although rudimentary in decor, they effectively present a kind of secret bureaucratic pedestrian walkway, not unlike Chicago's Pedway.

What the future holds

Today, several new projects are underway in the city, including new underground stations and sections and a proposed new Regional Connector (linking several existing lines), which will finally get to serve the oft teased Bunker Hill area with a new subway station.

In what must be the most extreme reversal of fortunes for any big city, LA is now blessed with not one urban rail expansion project, but two. Billionaire Elon Musk, whose Boring Company is already developing fascinating concepts for a Hyperloop ultra-rapid transit system to whisk people in pods rapidly over vast distances, recently announced plans for a slower, more localized network in LA itself. The first route, to be used for the initial testing, would be a 10.5-km (6.5-mile) 'tunnel' parallel to the 405 Freeway. The company has already applied for several route permits.

BELOW One of the most beautifully and creatively decorated LA Metro stations: Hollywood/Highland on the Red Line. Opened in 2000, designer Sheila Klein called the station architecture 'Underground Girl'.

Metres

0
— Belmont Tunnel
-10 — Boring Company test tunnel

-20

— Metro Purple Line extension
-30
— Grand Av Arts/Bunker Hill Station
(deepest Regional Connector station)
-40 — Regional Connector tunnels (at their
deepest point)

-50

-60

-70

-80

-90

-100

Hollywood/Vine Station and Boring Company Test Tunnel

LA does not spring to mind as having the most exciting underground features, and it is true to say that it was a late developer; but in recent years much has been added to the sub-scape. Although the 1925 subway terminal was ahead of its time, it was short and short-lived. Partly the result of Californians' dedication to personal motor vehicles, it took the city many years to plan a proper subway system. When it arrived in the 1990s, the architecture and design alone made it forgivably worth the wait. Since then many more structures have been commissioned below the surface.

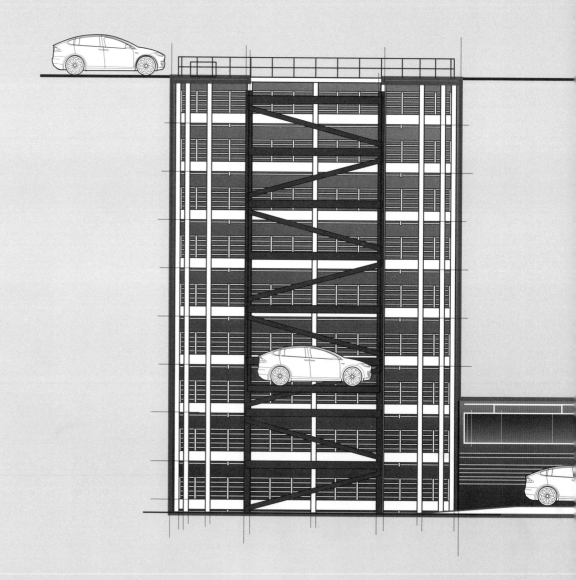

LEFT Along with the dozen other Red Line (now Line B) subway stations, the decor of Hollywood/Vine is exemplary. Local artist Gilbert Luján (aka Magu) chose light as the theme to reflect the neighbouring movie industry depicting the luminance of film projectors, the southern California sunshine and the stars of Hollywood. The station features two antique Paramount Pictures movie projectors and a roof space festooned with empty film reel containers. The pillars in this artist's impression are palm trees and a 'Yellow Brick Road' takes passengers to the surface.

BELOW Alongside ideas to place magnetically levitated pods in vacuum sealed tubes and shoot them at the speed of sound over very long distances (a concept called Hyperloop), billionaire Elon Musk is trialling a slower idea. His Boring Company (formed in 2017 as a subsidiary of SpaceX) opened a mile-long test tunnel in Hawthorne, California in 2018 to transport cars on metal 'skates' or sledges at up to 100mph. Elevators take car and passenger down to the tunnels and if successful the test site may be extended to Westwood. Another between LAX and Culver City could potentially cut a forty-five minute surface road trip to just five minutes under the ground.

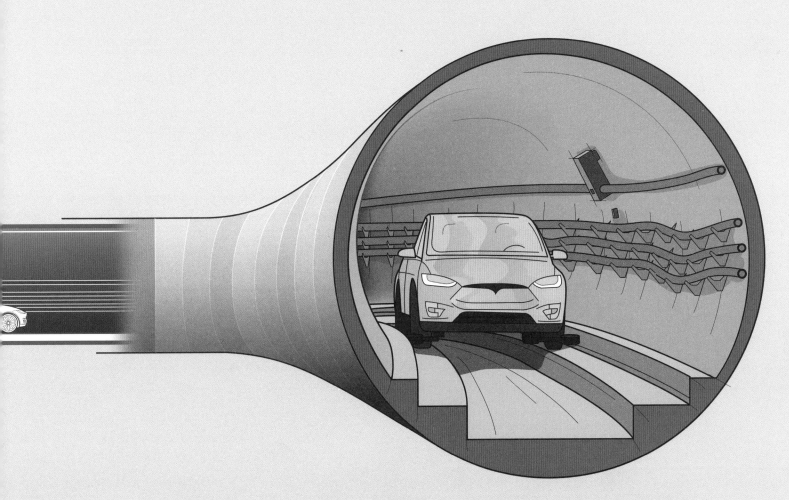

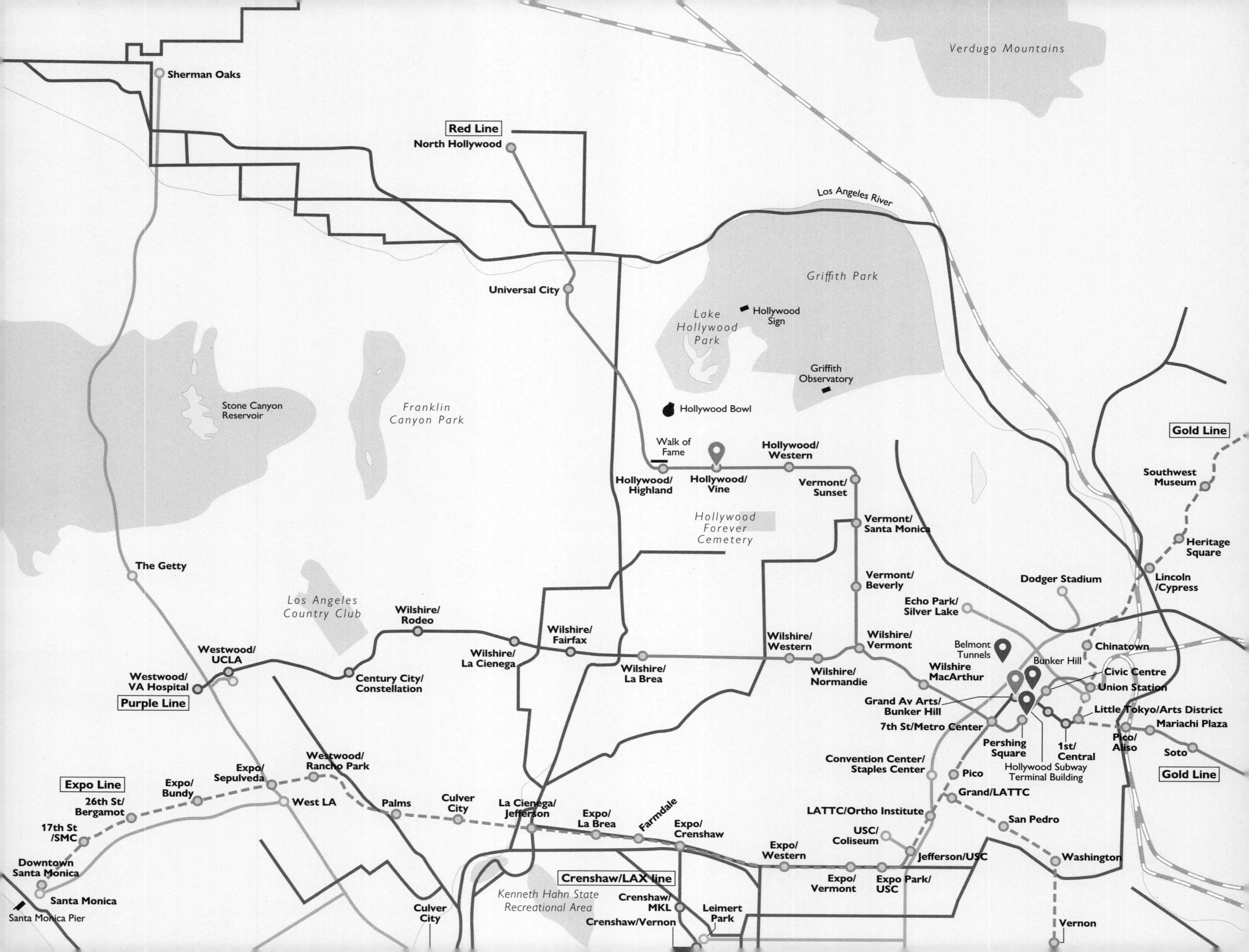

Transit Map – Los Angeles

Lines:
- Gold Line
- Red Line
- Purple Line
- Expo Line
- Crenshaw/LAX line

Stations:

Gold Line (north): Southwest Museum, Heritage Square, Lincoln/Cypress, Chinatown, Union Station

Gold Line (east/south): Little Tokyo/Arts District, Mariachi Plaza, Soto, Pico/Aliso, 1st/Central, Hollywood Subway Terminal Building, Pico, San Pedro, Washington, Vernon

Red/Purple Line: North Hollywood, Universal City, Hollywood/Highland, Hollywood/Vine, Hollywood/Western, Vermont/Sunset, Vermont/Santa Monica, Vermont/Beverly, Wilshire/Vermont, Wilshire/Normandie, Wilshire/Western, Wilshire/La Brea, Wilshire/Fairfax, Wilshire/La Cienega, Wilshire/Rodeo, Century City/Constellation, Westwood/UCLA, Westwood/VA Hospital, Westlake/MacArthur, 7th St/Metro Center, Civic Centre, Bunker Hill, Grand Av Arts/Bunker Hill, Pershing Square

Expo Line: 7th St/Metro Center, Pico, Grand/LATTC, San Pedro, LATTC/Ortho Institute, Jefferson/USC, Expo Park/USC, Expo/Vermont, USC/Coliseum, Expo/Western, Expo/Crenshaw, Farmdale, Expo/La Brea, La Cienega/Jefferson, Culver City, Palms, Westwood/Rancho Park, West LA, Expo/Sepulveda, Expo/Bundy, 26th St/Bergamot, 17th St/SMC, Downtown Santa Monica

Crenshaw/LAX line: Expo/Crenshaw, Martin Luther King Jr, Leimert Park, Crenshaw/MLK, Crenshaw/Vernon

Landmarks:
- Verdugo Mountains
- Los Angeles River
- Griffith Park
- Hollywood Sign
- Griffith Observatory
- Lake Hollywood Park
- Hollywood Bowl
- Walk of Fame
- Hollywood Forever Cemetery
- Echo Park/Silver Lake
- Dodger Stadium
- Belmont Tunnels
- Convention Center/Staples Center
- Franklin Canyon Park
- Los Angeles Country Club
- Stone Canyon Reservoir
- The Getty
- Kenneth Hahn State Recreational Area
- Santa Monica Pier
- Sherman Oaks

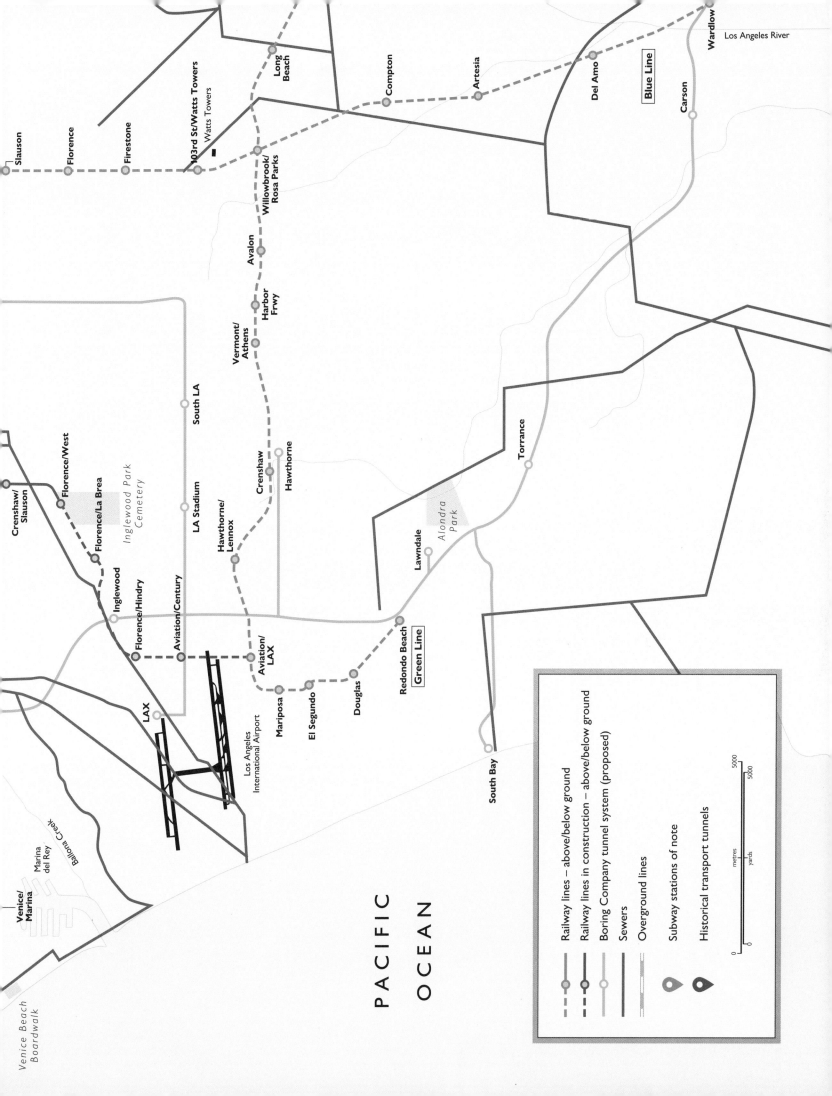

Mexico City

The city in the dried up lake

Founded by the Aztecs as Tenochtitlan around 1325, what is now Mexico City is the oldest capital in the Americas. Situated on a high plateau in the centre of Mexico, it has nearly nine million inhabitants and around twelve million more in the surrounding metropolitan area, giving it a population count larger than any other North American urban area. In 2016, the former Federal District was elevated to Ciudad de México status (shortened to CDMX), giving the city a level of autonomy equal to a state within the country.

Mexico City is used to being large. Even during the Aztec period it's thought that as many as 200,000 people lived in the vicinity by the time the Spanish arrived in 1519. The city was built on an island in Lake Texcoco. The colonists besieged and razed the city, but expedition leader Hernán Cortés recognized the importance of the site and founded modern Mexico City there, centred on the Plaza de la Constitución, known locally as Zócalo.

The Spanish conquerors deliberately developed the city as a capital for their empire in South America, so hundreds of grandiose colonial buildings, including many palaces, were constructed right into the nineteenth century – some with mysterious secret underground passages beneath them.

Although the lake around the city was once up to 150m (492ft) deep, it had been drying up since the Ice Age. The Aztecs controlled it with various dams and causeways. After several floods in the early 1600s, the new settlers began draining it entirely by making a 12-km (7.5-mile) ditch and tunnel to the Pánuco River. Periodic inundations continued, however, so a modern drainage system was finally constructed in the 1960s, using a huge network of several hundred kilometres of tunnels between 30m and 250m (98–820ft) deep. The unfortunate long-term consequences of this include water shortages, valleys becoming semi-arid, soil liquefaction during earthquakes and land levels having dropped by 10m (39ft) during the last hundred years.

Sewage solutions

With such a large population sanitation poses inevitable problems. In 2008 Conagua, the National Water Commission, started a mega-project to build a very wide and deep tunnel to drain waste from the city. The 62-km (38.5-mile) Tunel Emisor Oriente (TEO) should have taken four years to build at a cost of eleven billion Mexican pesos. It took two years longer and cost an extra four billion before it was ready in 2014. But, at 200m (656ft) below the streets and capable of carrying 150m^3 (5,300ft^3) of discharge per second, it's thought to be worth the wait – and the money.

Early transportation

To help move the populace around what was a rapidly expanding city in the nineteenth century, a horse-drawn street railway opened in 1858, running from Zócalo to Tacubaya. It was followed later that year by the arrival of a mainline railway

ABOVE At 8.7m (28.5ft) in diameter, the controversial Tunel Emisor Oriente (or Eastern Discharge Tunnel) runs up to 200m (656ft) beneath Mexico City and took four years to build.

to Villa Guadalupe. Expansion of both forms of transport continued sporadically so that, by 1890, Compañía de Ferrocarriles del Distrito Federal was operating 600 cars on 200km (124 miles) of route all pulled by 3,000 mules. In the following decade some of these were converted to electric power, being taken over in 1901 by the Compañía de Tranvías Eléctricos de México. The streetcars became the staple form of transport for city dwellers for five decades, but were falling out of favour by the 1960s as cars began to dominate the streets. By 1976 the streetcar network was down to just 156km (97 miles) of route on three lines, and the last one (between Xochimilco and Tasqueña) was converted to a modern light-rail line between 1986 and 1988.

Going underground

While a few ideas for urban transit had emerged in the decades before, it was not until the 1960s, as central streets clogged with cars, that Bernardo Quintana, founder of the engineering firm Ingenieros Civiles Asociados (ICA), produced a proposal that would lead to the Mexico City Metro. His broad concept of a three-line network was adopted as the Plan Maestro in 1967. The first line was built with record-breaking speed (about 1km/0.6 miles a month) so that 12.7km (7.9 miles) serving sixteen stations was ready to open in 1969. It adopted French rubber-tyred (*pneu*) train technology and ran initially between Insurgentes and Zaragoza stations.

By 1972, the network had been extended several times to serve forty-eight stations over a combined distance of 41km (25.5 miles). By 1985 the initially planned three lines had been expanded to their current lengths, and new Lines 6 and 7 had been added. Now, with almost 200 stations (115 of them underground, some sections up to 35m/115ft deep) the Mexico City Metro is the

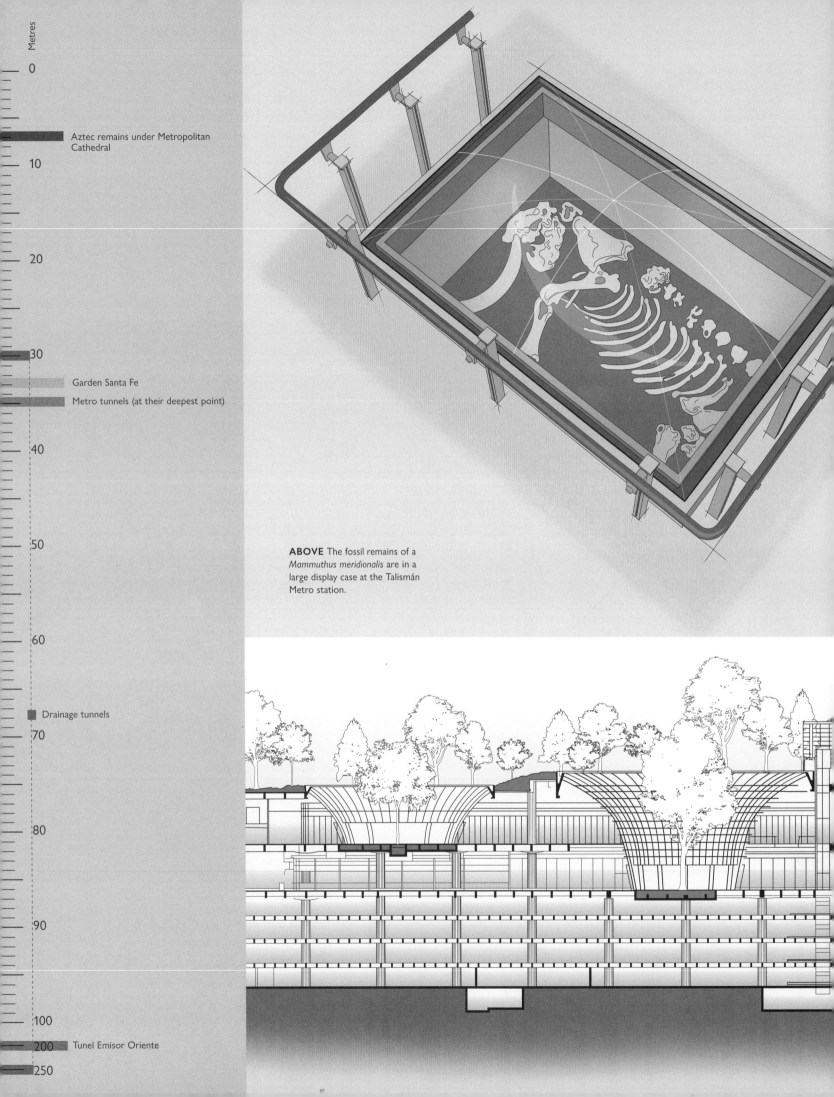

ABOVE The fossil remains of *Mammuthus meridionalis* are in a large display case at the Talismán Metro station.

Mammoth Bones at Talismán Station and Garden Santa Fe

Despite Mexico's susceptibility to earthquakes, tunnelling under its most unstable capital has not been deterred. The Metro is now the second biggest in North America, despite only opening its first line in 1969 – almost seven decades after the continent's biggest system in New York City. A total of 115 of its 195 stations are fully under the ground and archaeological discoveries have been unearthed as the system has expanded. During construction of Line 1 an Aztec idol and the bones of a Mammoth were found. In more recent years proposals for subterranean structures have abounded. Of all the fanciful underground proposals, the Garden Santa Fe is a beacon of what is possible.

BELOW Looking down from the skyscrapers that surround it, the shopping mall Garden Santa Fe almost seems like an organic spaceship. Located in the Central Business District, it was designed by KMD Architects and opened in 2014.

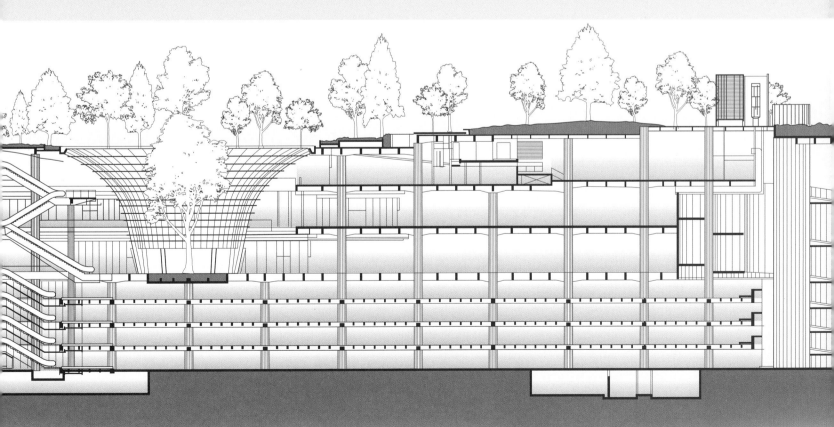

second largest in North America, topped only by that in New York City, and delivers 1.6 billion passenger journeys annually.

During the course of Metro construction over several decades, more than 20,000 archaeological finds have been unearthed. Everything from tiny objects to building foundations and from goddess statutes to pyramid blocks has been retrieved during the works, dating from prehistoric times to more recent Aztec history. Many are on display at the Instituto Nacional de Antropología e Historia (the National Institute of Anthropology and History). Some of the findings were so distinctive that they feature in pictograms designed for each station – a graphic system originally conceived of by American artist Lance Wyman.

The most iconic discovery lies in the remains of a woolly mammoth – a creature associated with much colder climates – the emblem of which adorns the Line 4 Talismán station built on the site where it was found. The mammoth itself has been reconstructed as an art exhibit inside the station.

Structural wonders

A shopping mall opened in 2014, and while there is nothing unusual about that, the Garden Santa Fe descends four stories underground and is full of trees! Situated at the heart of the financial district, the 12,000m^2 (130,000ft^2) site houses ninety stores and is up to 35m (115ft) deep. It's the first green underground mall of its type in the Americas.

Buildings going up – or down – in Mexico City need to be earthquake resistant. As testimony to the city's ability to achieve this, the 1985 quake registered 8.1 on the Richter scale, but most modern buildings, including the Metro stations withstood it well.

The city now looks set to make a new record in underground achievements. A groundscraper is

OPPOSITE During construction of the first Metro line in 1967, the remains of an altar dedicated to Aztec god Ehécatl were discovered and preserved. It is now displayed in the connection corridors between Lines 1 and 2 at Pino Suárez station.

BELOW The remnants of an Aztec temple dedicated to Ehécatl – the god of wind – which was built around the turn of the sixteenth century, was discovered in 2017. Archaeologists have concluded that the temple included some kind of sporting area, possibly a court for ball games.

a vast surface building extending a long way horizontally, but Mexico City could be adding a new word to the language: earthscraper. A building of sixty-five storeys and 300m (980ft) 'tall' was suggested during a competition in 2009 to be built from ground level *down*. It was designed by Esteban Suárez, founder of Bunker Arquitectura, in response to local laws preventing city-centre buildings exceeding eight stories in height. Suárez's proposal featured a Metro line running right through the heart of the building, with a station appearing as if suspended in the 'air'. Intended for the city's main square, the Zócalo, where it would cover more than 240m² (2,585ft²) at the surface, the building would contain homes, offices and a cultural centre or museum. Its upturned pyramidal-shaped structure was controversial to say the least. But it was not the shape that proved contentious — thanks to the Aztecs, pyramids are part of local culture. Instead, the biggest concern was seismic activity — both in terms of the effect a major tremble would have on such a building, and in terms of the potential for its construction to upset the delicate balance of forces below ground. Suárez's earthscraper was not the first to be proposed: a thirty-five-storey cylindrical version was suggested for Tokyo as far back as 1931. If built, however, the Mexico City earthscraper would be the deepest and most influential ever achieved. Since the competition, the restriction on building height has been lifted, and — for the time being, at least — Mexico City has settled on a shiny horizon of more traditional skyscrapers.

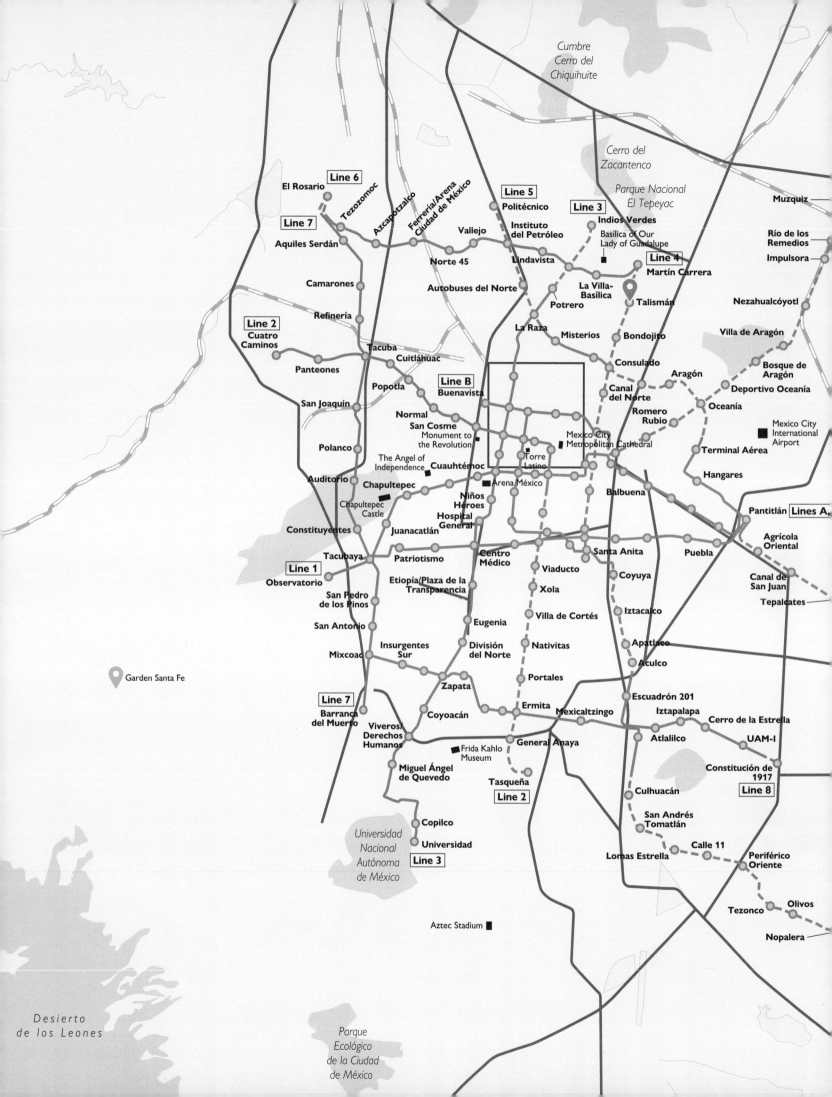

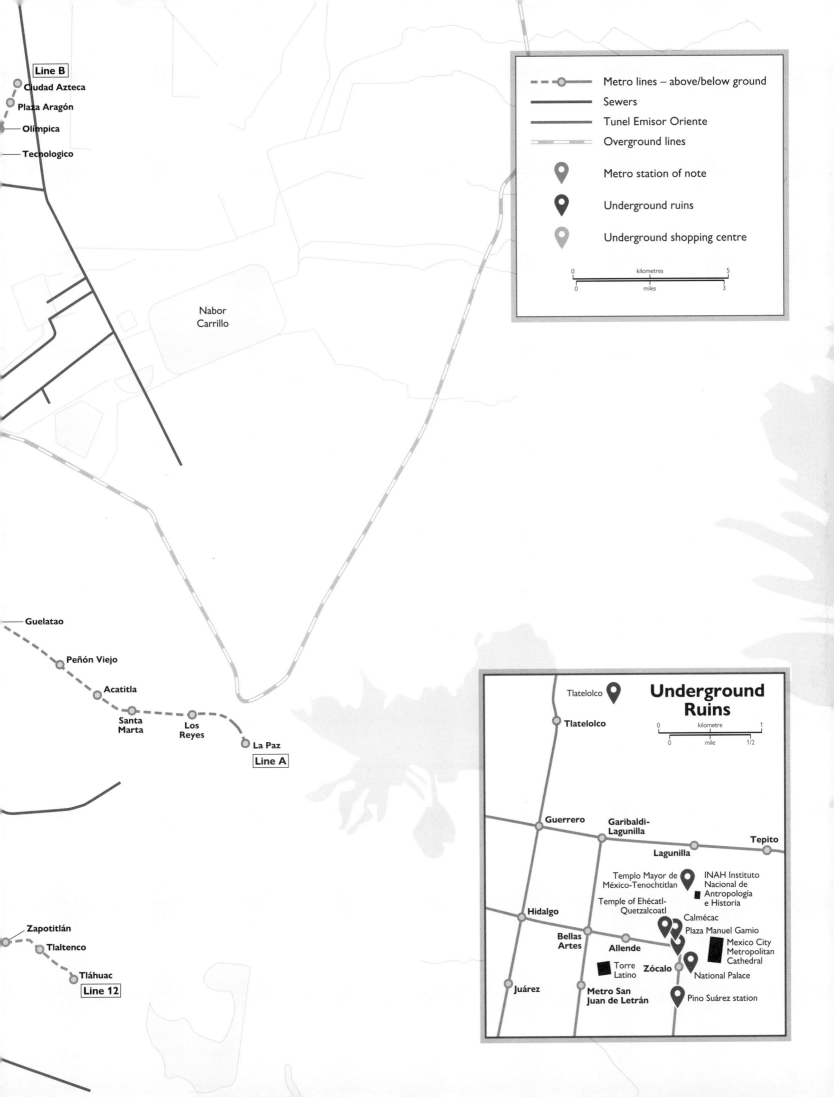

Chicago

The elevated city

The American Midwest city of Chicago has a population of of 2.7 million, with ten million in the wider metropolitan area. Looking at its orderly downtown district today, it's difficult to imagine the city's unplanned and mismanaged early years.

Elevating the city

Chicago expanded rapidly on the wide, flat and low-lying shores of Lake Michigan during the first half of the nineteenth century. At the time, its major thoroughfares (State, Dearborn and Madison Streets) all had completely uneven pavements and pedestrians were expected to walk up and down steps and precarious slopes just to proceed in a straight line along pathways paved with wooden slats, not stone. During the wettest periods, the water table in the soggy ground flooded the lowest-lying planks, which were left floating on a quagmire of polluted, sewage-filled water. Epidemics such as typhoid and dysentery were common – in one year, 1854, such diseases wiped out 6 per cent of the entire population.

Other than knocking down existing buildings and starting again, the only way to add proper sanitation and provide an even surface 'at grade' – that is, at pavement level – was literally to lift the level of the entire city. From the mid-1850s and continuing for around a decade, the biggest civil-engineering project of the day raised the height of central Chicago's sidewalks, *and* its buildings, by almost 2m (6.6ft).

Starting in 1856, civil engineer Ellis S. Chesbrough and his team began by building a new sewage system on the surface of the existing roads, which they then covered over to form a new grade level, with pavements on top of the buried sewage system. The buildings were then lifted upwards using an ingenious system of hydraulics and jacks so that their old entrances matched the height of the new road surface.

From 1858 onwards, hundreds of individual buildings, and entire city blocks, were jacked up with very few damaged, creating scores of new deep basements in the process. Within just a few years, engineers realized that the initial lift may not have been high enough in the event of Lake Michigan overflowing, so a second lift – this one taking some buildings almost 5m (16.5ft) above their original starting point – was proposed. Considered one lift too far for the city, the plan was not adopted. Had it been, Chicago might have built underground railways sooner than it did, given all the space that such an undertaking would have created beneath the city's streets.

Tunelling beneath the river

The Chicago River posed a major obstruction in the expansion of the city because it had to stay navigable. Bridge-building was limited, therefore, and any crossings that were built needed to be raised to allow ships through. Once settlements on Chicago's West Side had expanded to the extent that the bridges became congested, it was necessary to traverse the river in other ways. In 1867 construction began on a 490-m (1,600-ft) tunnel below the river between Clinton Street and

TOP 1857 illustration of the Raising of Lake Street on hundreds of jackscrews. An entire city block of 98m (320ft) was elevated in one go.

BOTTOM Designed to bring clean water directly into the city from further out into Lake Michigan, the Chicago Avenue water tunnel is inspected in this 1928 photo. The tunnel led to a collection inlet built several miles out into the lake. These structures, known as 'cribs', enabled Chicagoans to be supplied with much cleaner water than was previously available. The first of these was built 3.22km (2 miles) offshore in 1865.

Franklin Street West. Two years later the Washington Street Tunnel opened for pedestrians and horse-drawn carriages. Owing to initially poor construction the tunnel leaked and needed to be closed in 1884. A second under-river crossing was built between Michigan Street and Randolph Street North in 1871. The 576-m (1,890-ft) LaSalle Street Tunnel also suffered from poor construction, although it proved handy during the Great Chicago Fire of that year: as the flames tore through the city, leaving more than 100,000 residents homeless, hundreds fleeing the flames used the tunnel as an escape route.

Chicago's cable cars

Another issue those early city planners failed to address properly was that of mass transportation. The state intervened in 1859, by incorporating the Chicago City Railway (CCRy) and the North Chicago Street Railroad (NCSR) to run horse-drawn streetcar services on the new roads that been created by the sewage schemes. These two

Metres

0
— Original level of central streets
— PedWay

10
— L tunnels (average)
— Clark/Lake Station/Freight train tunnels

20
— Washington Street Tunnel/
Van Buren and Jackson Street Tunnel/
LaSalle Street Tunnel

Clark/Lake Station and Freight Train Tunnel

Outside of New York, Chicago has the most varied sub-surface infrastructure of any North American city – but oddly not so much of it is in subways. The two main subways here opened relatively late in comparison to other world cities with similar population sizes. What Chicago did have though was miles of freight-only underground railways, pneumatic post, water supply tunnels and basements created by raising street levels.

111 — Riverdale Calumet TARP Pumping Station

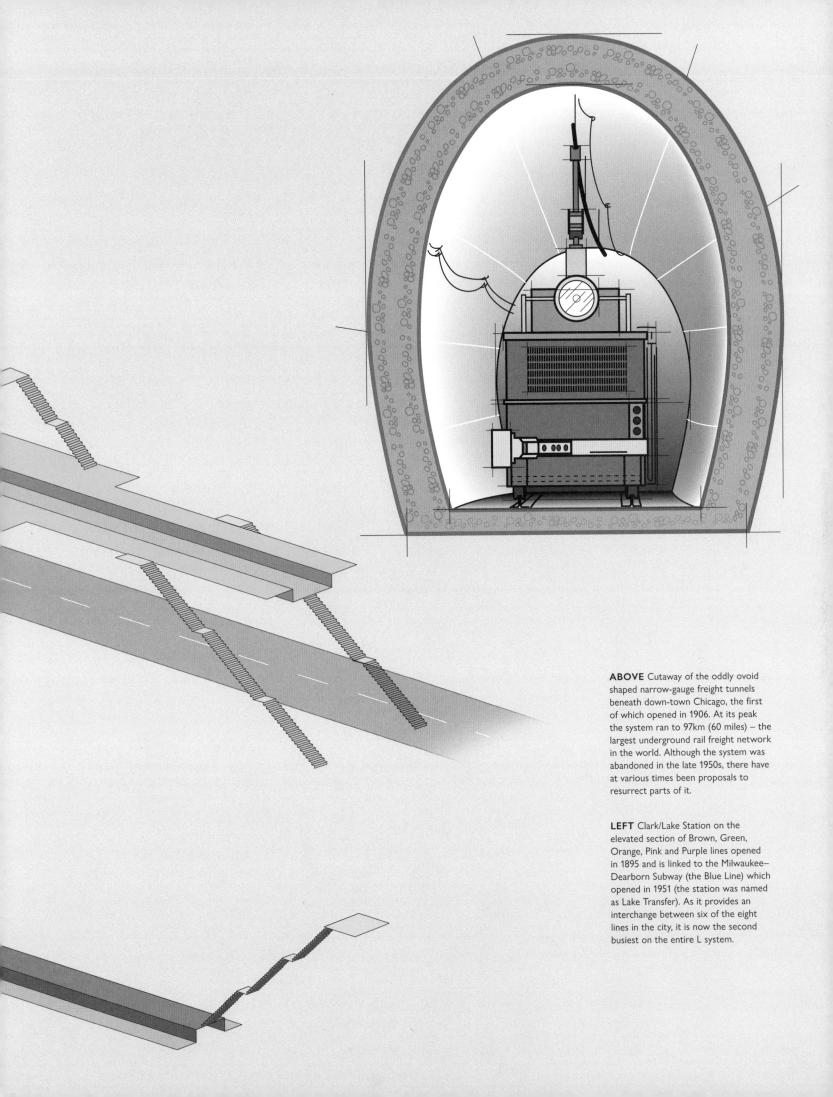

ABOVE Cutaway of the oddly ovoid shaped narrow-gauge freight tunnels beneath down-town Chicago, the first of which opened in 1906. At its peak the system ran to 97km (60 miles) – the largest underground rail freight network in the world. Although the system was abandoned in the late 1950s, there have at various times been proposals to resurrect parts of it.

LEFT Clark/Lake Station on the elevated section of Brown, Green, Orange, Pink and Purple lines opened in 1895 and is linked to the Milwaukee–Dearborn Subway (the Blue Line) which opened in 1951 (the station was named as Lake Transfer). As it provides an interchange between six of the eight lines in the city, it is now the second busiest on the entire L system.

ABOVE Running beneath the city's grid of perpendicular streets, the Chicago Tunnel Company's multiple freight lines had many junctions like this, photographed in 1924.

companies were joined by another in 1861, the Chicago West Division Railway (WCSR) – all serving different parts of the city, mostly separated by the Chicago River.

By 1882 the CCRy began using the latest motive technology to propel the streetcars, which involved installing a long moving cable in a road gulley. The cars clamped onto it in order to be pulled along the tracks. The other two companies followed suit, both opening cable-car lines of their own. By 1900 Chicago had the second-largest cable-car network in the United States, with more than 66km (41 miles) of cable. The NCSR even took over LaSalle Street Tunnel and refurbished it to accommodate cable cars that could pass under the river to serve West Side. The WCSR did the same in the old Washington Street Tunnel, and yet another tunnel was dug at Van Buren and Jackson Street in 1894. Sitting up to 18m (59ft) below ground, the new tunnels were deep, which meant they had steeper grades on entry and exit. Their 12 per cent inclines were almost three times the rise of many modern trains. After 1906, all three tunnels were converted to allow electrically powered street cars to use them.

Despite an intense network of electric and cable-hauled streetcars (amalgamated under Chicago Surface Lines in 1913), both fell out of favour and were replaced by the elevated railway. The tunnels were abandoned and sealed.

Chicago's rail innovations

Owing to the groundwater issue, and despite advances in other modes of transit and some tunnelling for streetcars and freight, Chicago was late in developing a full-scale subway system. The majority of the earliest public railways were raised on wooden and metal stilts giving rise, in 1892, to the Chicago and South Side Rapid Transit Railway,

ABOVE Speculative artist's impression on a postcard to promote Chicago's new subways being constructed beneath both main thoroughfares Dearborn and State Streets. This sketch of what the State Street Subway could resemble was published in 1941 showing a view looking north on State from Adams.

known colloquially as the Elevated (or 'L'). The system was still expanding as other cities were thinking of tearing them down and going underground. Elevated lines had been built, but largely removed, in New York (the first to open in 1868), Boston (1901) and Philadelphia (1922), and in Liverpool in the United Kingdom (1893), but only Chicago has stuck with them until today. The network is unique in having a 2.9-km (1.8-mile) 'loop', and the entire system now extends to 164km (102 miles).

Another unique feature of the city is Chicago's vast network of railed freight tunnels. Several tunnels had been conceived initially to house telephone cables. Just before construction began in 1899, the decision was taken to make the tunnels big enough to lay rails for specially designed freight trains as well as the cables. During the next seven years, the company excavated almost 100km (62 miles) of tunnel, up to 2m (6.5ft) high and around 12m (39ft) beneath Chicago. They were basket-handle (or horseshoe) shaped – similar to the Paris Métro tunnels. Miniature goods trains began carrying coal and parcels along the concrete-lined tunnels in 1906, helping supply the biggest users of fuel, such as City Hall or the huge Tribune Tower, and keeping dozens of wagons off the roads within the concentrated downtown area, particularly from 15th Street to Illinois Street and between Michigan Avenue and the rail depots in West Side. The freight tunnels were a real boost to the city's economy and were in use right up until 1959. Most of these tunnels remain intact, but the system was badly flooded in 1992 when construction pilings beside the Chicago River pierced the tunnels below.

Subways stayed in the planners' minds. In 1909, what became known as the Burnham Plan proposed 'an underground street-car system' along with new parklands, the widening of some streets, improved rail services, new harbours and substantial civic buildings. When it became technically feasible to ensure watertight seals and boring of tunnels, an entire subway network was planned. However, it was not until 1943 that the State Street Subway opened, having been built using a newly developed system of pressurized air to avoid the water bursting in. But its delay made the city fairly late to the underground railway club. Dearborn Subway followed just after the end of the Second World War. Planners took the wise decision to make the most of their subway construction. Instead of narrowing the tunnel in between stations, as happens on most other systems around the world, a continuous thoroughfare was built beside the tracks in the tunnel, so that passengers and pedestrians could walk from one station to the next – a feature that is unique to Chicago.

The two subterranean lines have since been absorbed into the overall L system and trains that run on the elevated lines descend into both the subway tunnels and then back out again to elevated tracks on the other side of the city. No new subway sections are planned, although the Red Line will see a new section of elevated rail line between 95th/Dan Ryan and a new terminal station at 130th Street.

The Pedway

Chicago has invested heavily in underground heated walkways for pedestrians. Managed by the Chicago Department of Transportation (CDOT), a series of long passages and hallways connects the lower levels of more than fifty facilities within the central 'loop' area. The Downtown Pedestrian Walkway System (Pedway for short) began simply with connections between subway stations such as Washington, Lake and Jackson. Since 1951, other sections have made it

ABOVE 1940s postcard depicting a BMT-style 'Bluebird' car on Chicago's initial subway line.

TOP RIGHT Shops now fill most of the Pedway. Construction began in 1951 when the City of Chicago built one-block long tunnels to connect Red Line and Blue Line subways between Washington Street and Jackson Boulevard.

BOTTOM RIGHT In some locations connections are needed between different sections of the Pedway. Here a 'bridge' is located beneath Upper Randolph and above Lower Randolph Streets.

more about links to shops such as Macys, to hotels, skyscrapers in the Illinois Center Complex and municipal buildings that include the James R. Thompson Center and City Hall, and even to a few residential blocks – for example, the new Aqua tower. The Pedway can be entered on ground level, where the black/red/yellow compass logo is seen, especially along Michigan Avenue. As well as providing access to buildings above, the tunnels also house a number of facilities themselves, from coffee shops and watch repairers to barbers and shoeshiners. The idea is so successful that many other buildings are linked to each other underground, even though they are not part of the main Pedway network.

Deep, deep down

If the depths listed so far seem shallow, how about taking a trip that delves 111m (364ft) below the surface? The Metropolitan Water Board of Greater Chicago has built something far below the Riverdale neighborhood at 400 E. 130th St. It's the Calumet TARP Pumping Station and is the lowest anyone can get beneath the Windy City.

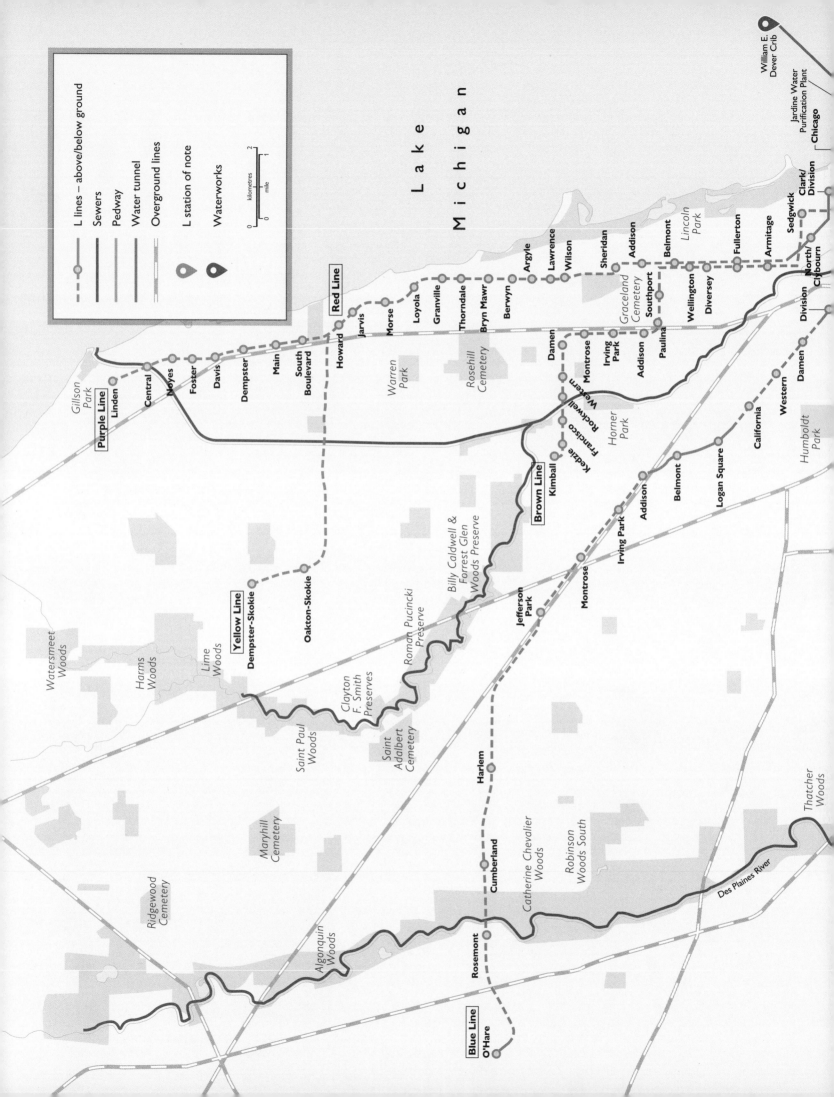

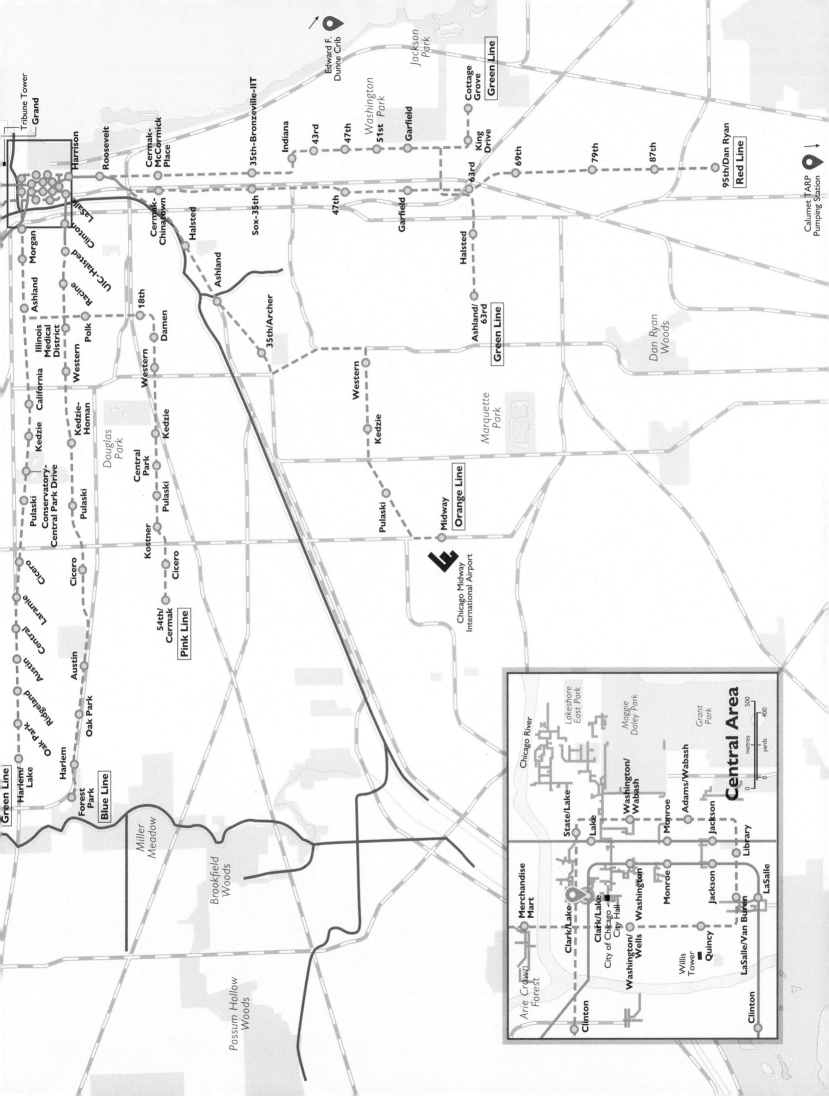

Cincinnati

The subway that never happened

In the southwest corner of Ohio, USA, in a flat basin surrounded by hills, on the banks of the river that shares its name with the state, Cincinnati is one of America's great historic cities. Now with a population of 300,000 and more than two million in its metropolitan area, it is the state's largest city. Founded in 1788, Cincinnati grew rapidly as a seat of culture and education, a major stopping point for steamboats and a centre for paper and type production. For a short period in mid-nineteenth century it was the sixth-biggest city in the country. Although other Midwestern giants soon eclipsed it, including Chicago, Cincinnati's fathers maintained lofty ambitions and it became home to the world's first reinforced concrete skyscraper – the sixteen-storey Ingalls Building – which opened in 1904.

Thwarted subway plans

Cincinnati's streetcar network opened with horse-drawn vehicles in 1859, followed by some cable-hauled and incline railways; it was electrified from 1889. By the time of its peak, in the mid-1920s, it had expanded to cover 357km (222 miles) of track and was carrying 100 million passengers per year. A victim of its own success, causing traffic congestion and accidents with early private vehicles, the streetcar network needed relief. As happened in other cities, the idea of an underground subway system was mooted.

An impressive waterway had been built in 1825, linking Lake Erie with the Ohio River, and with its opening into the river passing right through downtown Cincinnati. But the Miami and Erie Canal, as it was known, traversed hilly terrain and needed many locks, which made it slow going. The water also tended to freeze over in the harsh winter months. Being fairly late on the scene in artificial waterway terms (the more successful Erie Canal through New York state had opened much earlier), the canal soon lost the battle for transporting freight in fierce competition with the newfangled railroads.

The expensive canal became unprofitable by 1856 and unused by 1877, after which it was dubbed the 'dead old ditch'. It had been abandoned by around 1906 and, despite remedial waterworks after the Great Flood of 1913, which killed over 600 across Ohio when rivers burst their banks, the infrastructure of the canal was effectively destroyed, rendering it useless. Many parts of it were filled in and covered over.

Spotting a chance to save some money on the construction of a future subway system, Cincinnati Mayor Henry Thomas Hunt took control of the abandoned canal in 1910, with a view to including it in plans for a 26-km (16-mile) combined elevated and underground 'loop' around the suburbs that would utilize the canal bed in the downtown area. Between 1920 and 1925 around $6 million dollars were spent converting 3.5km (2.2 miles) of the old canal into a track bed for the subway. But costs rose and corners were cut, leading to poor construction and law suits for damage to buildings above. Work was halted by the stock market crash of 1929 and ensuing Great Depression, but not before 11km (6.8 miles) of tunnels and four

ABOVE Artist's impression of the never-used Cincinnati subway construction, using the cut'n'cover method.

underground station areas complete with platforms had been built. A new boulevard, Central Parkway, had also been created on top of the covered canal section. But no track had been laid below ground.

Several attempts were made to rekindle the subway concept. In 1936, 1939 and 1940 various ideas were put forward, but plans were mothballed in 1948 and scuppered in the 1950s when, in order to save money on digging a new water main, the pipes were simply placed in what would have been the northbound subway running tunnel. In the 1960s there was a plan to convert the proposed station site at Liberty Street into a nuclear fallout shelter. A wine cellar, nightclub and filming set were also proposed, none of which ever came to fruition. MetroMoves, a plan to introduce light rail in 2002, would have utilized some of the tunnelling but the scheme was voted down by 68 per cent of the public.

One project that has successfully utilized a small section of the underground space created is the Riverfront Transit Center, a multimodal hub for bus and coach services that opened in 2003. It would even one day have seen trains had the proposed 27-km (16.8-mile) Eastern Corridor Commuter Rail project been built. This concept involved major road and rail improvements for southwest Ohio. Among them, the Oasis Rail Transit project would have been an initial ten-station portion of a new regional rail system, with the line running between the Riverfront Transit Center in downtown Cincinnati to Milford. The project stalled in 2012, however.

Legacy of failure

Owing to its length and essentially solid construction, the bulk of the original underground network remains intact, enabling Cincinatti to gain a kind of legendary status among transit geeks as one of the most lamented unbuilt systems in existence. Many have cited the city's failure to complete the system as a partial cause of its long decline. The last section to have been built back in 1929, the Hopple Street Tunnel provides the best opportunity for urban explorers to gain entrance to the long empty voids, when two sealed-off portals are occasionally opened for official tours.

Toronto

Weatherproof shopping

Canada's largest city is located on the northwestern shore of Lake Ontario, the most easterly of the Great Lakes, and has 2.7 million inhabitants with about six million in the Greater Toronto area.

Archaeologists say that there have been people living on the shores of Lake Ontario for at least 10,000 years. When the Europeans arrived in the early 1700s they found Iroquois people living in small villages including one called Teiaiagon on the bank of the Humber River. French colonialists established fur trading posts at Fort Rouillé and Fort Toronto in the mid-1700s, but these were destroyed during the Seven Years' War. It was not until the late-eighteenth century that a new fort and settlement grew up nearby, which was named York. The small port of less than 10,000 was given city status in 1834, when the name was changed to Toronto in order to help distinguish it from other towns with the name York. It became a welcome home to immigrants. Industry and distilleries arrived in the early 1800s as did the railways, in 1853.

Early infrastructure

Downtown Toronto was quickly rebuilt after the Great Fire of 1904, which destroyed more than 100 buildings in the area. Not long after, attention turned to improving public transport. The city had already grown a very large streetcar network after their introduction in 1861 and lines were electrified from the 1890s onwards. But as the city expanded, private streetcar operators were less keen to reach further out, so the system fell under the control of a transit commissioner. Unlike many other cities, Toronto has held on to ten of its original streetcar routes which, now fully modernized using light-rail vehicles, are a vital part of the transit offering. The city has four streetcar stops in tunnels: the streetcar-to-subway interchanges at Spadina, St. Clair West and Union, plus a tunnelled station under Bay Street at Queen's Quay, which is for streetcars alone.

As in Paris, London and Chicago, Toronto also had a pneumatic mail system, but it was quite a bit smaller here: just 4.5km (2.8 miles) of tubes between two newspapers, City Hall, Union Station, the Royal York Hotel and the fifteenth storey (the top floor) of the Canadian Pacific Railway head offices. Although effective, the system did not spread beyond that.

Toronto Subway

Frustrated by the resistance of the private streetcar operators to work with them, in 1909, the city council started to look at ideas for building some new streetcar lines in tunnels. The idea went to a public vote and Torontonians supported the it. Initial plans for an 18-km (11-mile) network on three lines were drawn up a year later by a company called Jacobs and Davies. Costs scuppered the scheme, and despite later efforts to revive the idea, they all failed until a public referendum supported a new scheme in 1946. Even so it took three years to start construction of the first 7.4-km (4.6-mile) line along the busy thoroughfare of Yonge Street. Almost entirely underground, it was built using the cut-and-cover method, so the stations are relatively shallow.

A new line from Eglinton to Union opened in 1954, and was extended piecemeal during the coming decades. The Toronto Subway now has four lines covering almost 80km (50 miles) – 60km (37 miles) of it underground – and serving seventy-five stations.

The current system has a couple of ghost stations. The concept for an east–west subway beneath the long Queen Street goes back to 1942, when there was a proposal for two streetcar tunnels, the other running north–south between Bay and Yonge Streets. A metro-style solution was chosen for the north–south route (Yonge subway) but the lengthy Queen Street would continue to be served by its original streetcars. However, while construction was underway on the subway itself, an interchange station was carved out of the rock in case the east–west Queen subway ever materialized. Sometimes called Lower Queen station, (others refer to it as City Hall), the lifeless platforms lay buried and unserved directly below the current platforms of the existing Queen station. At various times since (up to 1966 at least) it has been included in plans for other new lines, but so far it remains empty.

One station that saw service for just six months was called Lower Bay. Situated on the Bloor-Danforth line (now Line 2), which opened in 1966, is a station in the Yorkville area named Bay. In an attempt to create the impression that every station on the network was served by two routes, the Transit Commission experimented with running trains in patterns that included portions of both the Bloor-Danforth and the Yonge lines. The resulting three-line service pattern meant that the platforms built immediately beneath the main Bay station (Lower Bay) were also served by alternate services. The idea was sadly flawed because it caused confusion with Bay and Lower Bay passengers waiting on the stairs

BELOW Vaughan Metropolitan Centre, which opened in 2017, is one of only two subway stations outside the Toronto city limits. The mirrored and coloured glass dome effect is an artwork called *Atmospheric Lense* by the Paul Raff Studio and the surface entrance (not shown) is a glass and steel encased ovoid by Grimshaw Architects.

Metres

0
— Spadina Storm Trunk Sewer
— Casa Loma tunnel

-10
— PATH

-20
— Lawrence Station (deep subway station)

-30
— Copeland transformer station tunnel
— Avenue Station (deepest light rail station)
— Scotia Plaza gold bullion vault

-40

-50
— Western Beaches tunnel

-60

-70

-80

-90

-100

PATH and The Vault

Although Toronto was the first Canadian city to open a subway line, its development after that faltered and Montreal (a smaller city) overtook it to have the most underground track. However, with the recent opening of the 8.6-km (5.3-mile) Line 1 extension to Vaughan, and several other projects under way, (including the major crosstown Line 5), Toronto could soon regain the country's subway crown. The two cities also run neck-and-neck for the length of their underground connected shopping malls. Toronto is also home to number of impressive safes and vaults.

BELOW Artist's impression of The Vault. One of the quirkiest venues for a corporate function, it was first built as the nation's bullion reserve. The 420m^2 (4,500ft^2) space below 250 University Avenue was built with thick concrete walls and a 1-m (3-ft) deep vault door. It was completely refurbished with modern catering facilities and is now full of leather furniture, bars, and even a putting green! There is another large gold vault beneath Scotia Plaza but this too has ceased being used for its original precious purpose.

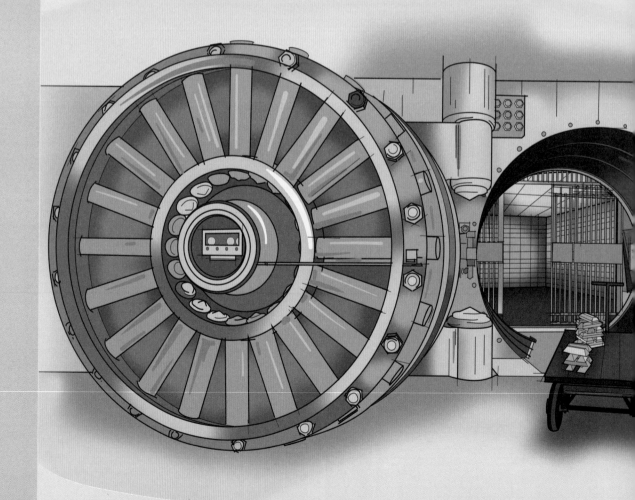

ABOVE One of the most successful and architecturally pleasing sections of the labyrinthine Toronto PATH is the Eaton Centre. Named after one of the country's leading retailers, Eaton's department stores (which began life as a little dry-goods store in 1869), the company ceased trading in 1999. The Eaton Centre now has more than 230 shops over two city blocks and its operators claim it rivals the CN Tower for numbers of tourists.

ABOVE Named after the former commissioner of Toronto's public works, the R.C. Harris water treatment plant's art deco surface building is of great civic pride. The marble covered entrance ways and filtration halls have helped earn it the nickname the 'Palace of Purification'.

between the upper and lower levels so as not to miss the first available train. The tracks however remain in use for personnel training, and also double up as a film set and occasional concert venue without interfering with the running of the lines in service above.

Toronto's Line 3 Scarborough opened in 1985 as a 6.4-km (4-mile) automated light-metro shuttle, but is due to be converted to heavy-metro standards (as an extension of Line 2). The 5-km (3-mile) Line 4 Sheppard opened in 2002 and is entirely underground. Scheduled for extensions at either end, neither has yet been funded. However the brand new east–west crosstown route under construction, Line 5 Eglinton, will run 19km (12 miles) with at least a dozen underground stations. Officially designated as light rail it is scheduled to open in 2021. A proposed 11-km (7-mile) light-rail Line 6 is also on the drawing board and scheduled to open in 2023, but it would not have any tunnelled sections. Several other extensions are planned as well: the Bloor-Danforth subway is to be extended 6 km (3.7-miles) to Scarborough by 2026. The Yonge subway may one day gain five new stations to the north but this plan is as yet unfunded.

The regional commuter rail provider GO Transit also uses some tunnels in the city. Its Metrolinx RER programme includes laying 150km (93 miles) of new track, and some of this, for example under Highways 401 and 409, will be in tunnel.

Underground waterways

Toronto has numerous well-built sewers. Many started as small streams, such as the 7-km (4.3-mile) Garrison Creek in the west end. It was converted to a drain in the late 1800s by being culverted over. Rosedale Creek was converted in a similar way in 1888. Now known as North York Storm Trunk Sewer and Spadina Storm Trunk Sewer, the two were turned into art in 2013, when photographer Michael Cook exhibited images of them.

More than a dozen underground freshwater reservoirs keep the city supplied all year with drinking water. Some are topped by sports grounds, others are built over differently. The mid-2000s saw a secretive tunnelling project by a utility company seeking to use lake water as a coolant during the city's heatwaves. They bored 15m-diameter (49-ft) tunnels to pull the cold water inland for use as a natural coolant to help air-condition Toronto's many tall buildings.

Subterranean shopping

Like Montreal, Toronto has developed a huge network of underground passageways connecting shopping malls and major buildings. The PATH can trace its roots to 1900 when T. Eaton Co built an underground passageway to connect its main store on Yonge Street with a neighbouring budget store. In just a few years others had followed suit and, by 1917, there were at least five tunnels connecting buildings in the downtown area.

When the Union Station opened in 1927, the planners constructed a tunnel to a nearby hotel in order to expedite a weather-free passage between the two. In the 1970s a huge office development at Richmond-Adelaide was connected with a new hotel tower, the Sheraton Centre. From this point, the tunnels began to expand all over the city. In the 1980s a proper signage system was needed to help make the interlinking passages easier to navigate. PATH now connects seventy-five buildings and six subway stations and has more than 1,000 restaurants within almost 370m^2 (4,000ft^2) of retail space. Sales of almost two billion Canadian dollars are generated by the 200,000 people passing through every day.

Quirky spaces

The former psychiatric hospital at Lakeshore was built in a Gothic and Romanesque Revival style during 1884. Owing to its policy of dealing with mental illness using smaller units to house patients, several smaller buildings were also erected in the grounds. A secure method was needed to get the staff, equipment, supplies and patients from one building to another, so a series of tunnels was constructed beneath the buildings. Although the site is now occupied by a college, occasional tours provide a glimpse of the creepy tunnels.

The very grand Casa Loma looks like an ancient castle, but was built in 1914 in the Gothic Revival style for the financier Sir Henry Pellatt. A 250m (820ft) tunnel connects the main castle to its hunting lodge and the stables.

Canada is one of the worlds leading sources of gold, and way below the PATH and the subway, is the deepest room in Toronto – it's a 400m^2 (4,300ft^2) gold bullion vault under the Scotia Plaza skyscraper in the heart of the Financial District.

At Bremner, on the site of a former railway shed, the water utility company Toronto Hydro is building a new transformer station underground. The old steam loco roundhouse will then be reconstructed above it, brick by brick. The 200 million Canadian dollar project is needed as the downtown area has seen such a massive boost in population and shopping malls, all requiring additional water.

BELOW With ninety-eight rooms, the Casa Loma was once the largest private residence in Canada. Its gothic revival style cost owners $3.5 million and took three years to construct – opening in 1914. A tunnel between the house and stables has now been opened for public visits as 'The Dark Side Tunnel Exhibit'.

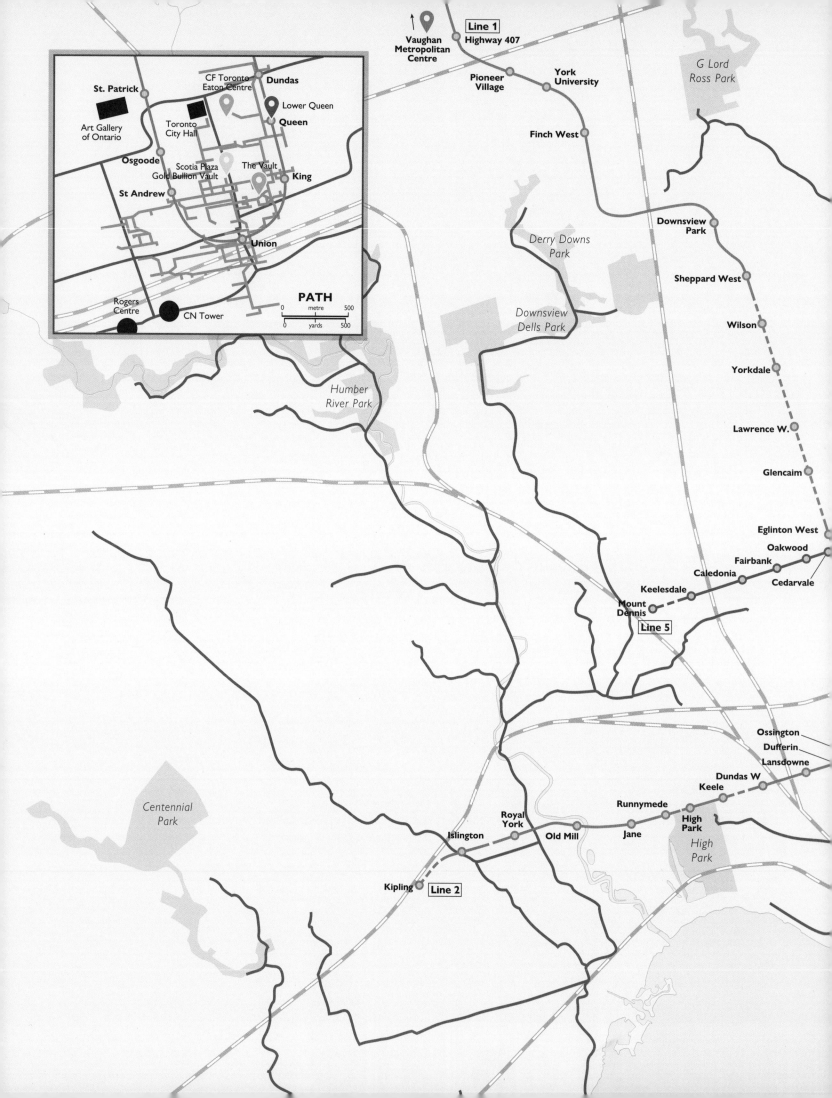

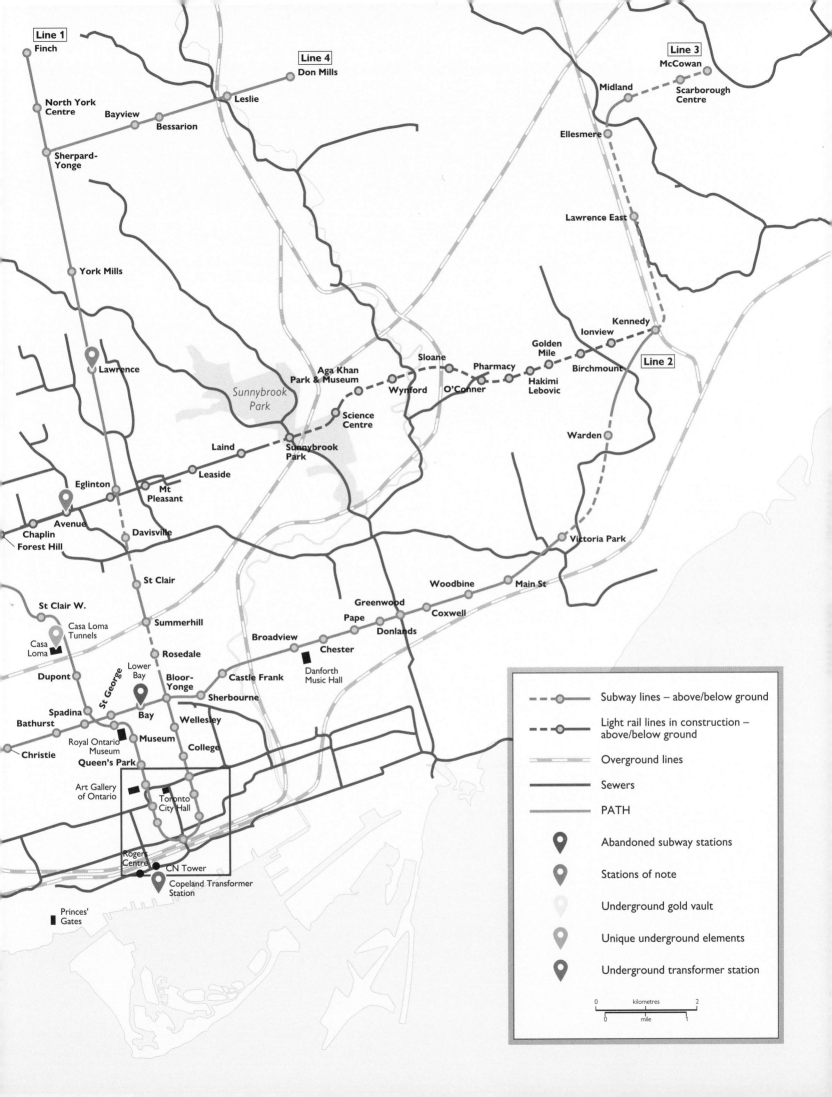

Montreal

Bilingual basements

The largest city and economic centre of the Canadian province of Quebec sits on an island where the Saint Lawrence and Ottawa rivers meet. At almost 500km² (193sq mi), the Île de Montreal is five times the size of Paris and could accommodate eight Manhattans – so it's big! In the centre are the three peaks of an ancient volcanic hill, the Mont Royal. Alongside its volcanic past, 15,000 years ago during the Ice Age, the weight of glaciers caused fissures in the rock beneath them. One of these, the Saint-Léonard Cavern, was discovered in 1812 and, recently, another huge void was discovered leading from it and underneath the Parc Pie-XII in the city's eastern district. It is up to 250m (820ft) long, 6m (19.7ft) deep and filled up to a depth 5m (16.5ft) with water.

Around Montreal's peaks, urbanization covers much of the island, housing a population of two million, with nearly 500,000 on neighbouring islands and another 1.5 million in the wider metropolitan area. Montreal is the second-largest French-speaking city in the world after Paris.

From village to city

There is evidence of settlement on the island up to 4,000 years ago and a village called Hochelaga at the foot of Mont Royal was established 200 years before the French arrived.

By 1611 explorer Samuel de Champlain established a fur trading post near where the present Pointe-à-Callière Museum of Archaeology stands in Old Montreal. Colonists arrived thirty years later, establishing Ville-Marie on the southern shore of the Île de Montréal in 1642. Despite many battles and attacks, the settlement grew. The opening of Lachine Canal in 1826, new bridges and the imminent arrival of railways in 1836 led to the incorporation of Montreal as a city in 1832. By 1860 it was the largest city in what was then known as British North America.

Subterranean waterworks

As the population grew, especially during the 1800s, so did its needs for water and sanitation. During this time – and as early as 1739 in some cases – many former rivers and creeks were diverted or covered over, including the Saint-Pierre River. By 1832 the William collector sewer had been built – a 350m (1,148ft) stone conduit in which part of the river was buried. A century later almost one-third of the river's entire length had been used for the sewage system, so that today just 200m (656ft) of the once substantial river remains on the surface in Montréal-Ouest. By the 1870s, almost every stream or creek in the downtown area had been covered over or diverted.

As new sections of the large Île de Montréal were urbanized, so yet more new waterworks were required. After the Second World War, plans were made to ensure that all newly built areas would be properly served by sewage and water systems. The concept was that the city region would grow to seven million inhabitants by the year 2000, although only half that figure has been reached. One such waterwork is the massive Meilleur-Atlantique collector sewer in Cartierville,

ABOVE At the Pointe-à-Callière museum there is a beautifully lit path through the old sewer collector.

which was built in 1953. Another is the Decarie Raimbault system, where tunnels lead through the limestone. There are also massive storm drains, — for example, on West Island — and underground aqueducts completing almost 5,000km (3,100 miles) of sewer and waterworks below Montreal.

Underground City

Montreal's humid continental climate provides fine warm summers, but the winters can be extremely harsh, with temperatures dropping to as low as −30°C (−22°F). Then there is the precipitation with an average of about 2.1m (6.9ft) of snowfall a year. In 1962, the building of a skyscraper called the Place Ville Marie was happening under the direction of urban planner Vincent Ponte. He conceived of a massive, expanding subterranean space, air-conditioned to avoid the extremes of weather on the surface and connected to other parts of the city. The tower and its basement shops were linked via tunnel to the Central Station and the Queen Elizabeth Hotel. The winning of Expo 67, the start of work on the Montreal Métro and securing the 1976 Summer Olympic Games, enabled further tunnelled connections to be extended to Bonaventure and Windsor stations, the Château Champlain hotel and Place du Canada office tower. So began the core of what Ponte described as the 'underground city', locally known as the RÉSO – based on the French word *réseau*, meaning 'network'.

The Métro system connected a further ten buildings directly with its stations, and as other major commercial developments began in the 1970s – for example, Complexe Desjardins, an office and retail complex in downtown Montreal

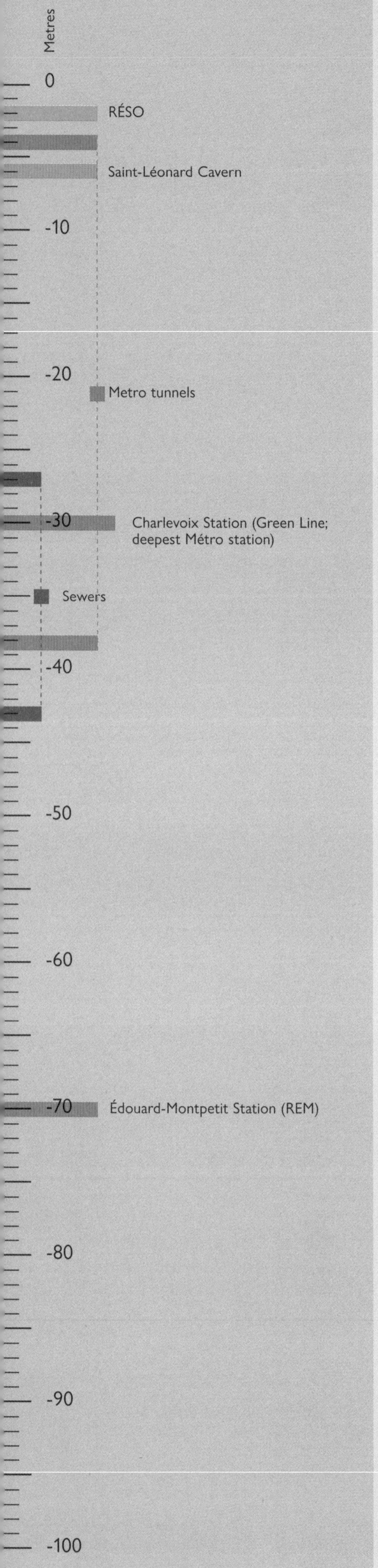

Metres

- 0
- RÉSO
- Saint-Léonard Cavern
- -10
- -20
- Metro tunnels
- -30 Charlevoix Station (Green Line; deepest Métro station)
- Sewers
- -40
- -50
- -60
- -70 Édouard-Montpetit Station (REM)
- -80
- -90
- -100

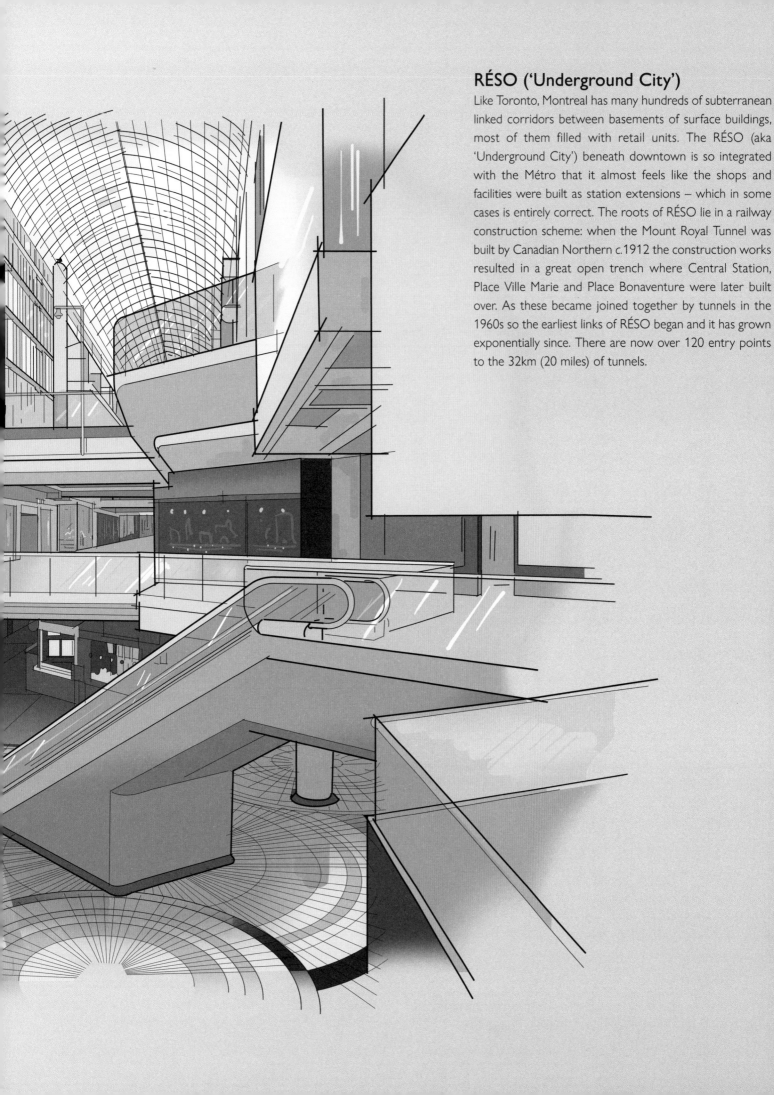

RÉSO ('Underground City')

Like Toronto, Montreal has many hundreds of subterranean linked corridors between basements of surface buildings, most of them filled with retail units. The RÉSO (aka 'Underground City') beneath downtown is so integrated with the Métro that it almost feels like the shops and facilities were built as station extensions – which in some cases is entirely correct. The roots of RÉSO lie in a railway construction scheme: when the Mount Royal Tunnel was built by Canadian Northern c.1912 the construction works resulted in a great open trench where Central Station, Place Ville Marie and Place Bonaventure were later built over. As these became joined together by tunnels in the 1960s so the earliest links of RÉSO began and it has grown exponentially since. There are now over 120 entry points to the 32km (20 miles) of tunnels.

– yet more connections arose. Between 1984 and 1992, the underground city expanded from 12km to 22km (7.5–13.7 miles) with the construction of three massive, linked, underground shopping malls. Major new office buildings and Métro extensions have seen the complex increase so much that it is now impossible to imagine a new downtown building not connected to the Underground City, and it is firmly established in the fabric of Montreal life.

Montreal's Métro

The idea of developing a metro system for Montreal goes right back to start of the twentieth century. The federal government even created the Montreal Subway Company in 1902 to promote the concept. From 1910, proposals were submitted by several independent and existing transit operators, including the streetcar company. But with no funding available and mainline railway objections, nothing progressed.

In the wake of the Great Depression, two world wars and traffic congestion, a new plan for a two-line transit network was produced in 1944. One line would run beneath the main thoroughfares of Saint-Jacques, Notre-Dame and Saint-Denis Streets and the other below Sainte-Catherine Street. The project stalled as streetcars were replaced by buses on the surface, but was then revised with a myriad proposed extensions that ultimately appeared too expensive. The map for a 1953 project looked suspiciously similar to the metro system being built in Toronto at the time. Still Montreal was no nearer to mass transit.

Finally, in 1961 the city council voted for a scheme based on a rubber-tyred wheel technology (*pneu*) similar to that used on some Paris Métro lines. Three routes were planned, but construction began in 1962 on the Line 1 (Green Line) and Line 2 (Orange Line), culminating in the opening of the Métro de Montréal between October 1966 and April 1967.

The short Yellow Line was subsequently bored beneath the river to the neighbouring city of Longueuil, with a stop serving the site of the Expo67 exhibition, bringing the early network to twenty-six stations on three lines.

The network has since been expanded piecemeal over the years, with the addition of a new Blue Line in 1986. It now has four lines serving sixty-eight stations and 70km (44 miles) of track. Although some of the 1970s proposals for a metro system might be described as exuberant, the idea of creating a 160-km (100-mile) tunnelled network for the six or seven million expected inhabitants was not so crazy. After several steps forward and back, plans are now advanced for an extension of the short Yellow Line further into the city of Longueuil and the Blue Line should reach its longed-for new terminal in Anjou in the next few years.

ABOVE Orange Line Métro station Namur opened in 1984 was designed by Labelle, Marchand et Geoffroy. The photo shows *Système*, an illuminated aluminium sculpture by artist Pierre Granche, suspended above the ticket hall.

A brand new route, known as the Ligne Rose (Pink Line) was proposed as part of the new Mayor of Montreal's electoral campaign in 2017.

A separate Réseau Électrique Métropolitain (REM), currently under construction, will bring fast, light-rail vehicles from underground stations in the downtown area to the Montréal-Trudeau airport in just twenty minutes, and is expected to open by 2021. Its new platforms below Édouard-Montpetit Métro station will be 70m (230ft) below ground – the equivalent of twenty storeys – making it the second-deepest station in North America. Although the majority of the REM will run on the surface in the suburbs, the central section will give new life to the 1911 Mount Royal Tunnel (Canada's third longest) which had all but become redundant in recent years.

BELOW Charlevoix station on the Green Line, opened in 1978, is unusual in that one platform is stacked on top of the other, the lower one being the deepest on the Métro network at 29.6m (97ft). It was designed by Ayotte et Bergeron with these beautiful stained glass artworks by Pierre Osterrath and Mario Merola. The Belgian based Osterrath stained-glass makers also contributed works to Berri-UQAM and Du College Metro stations.

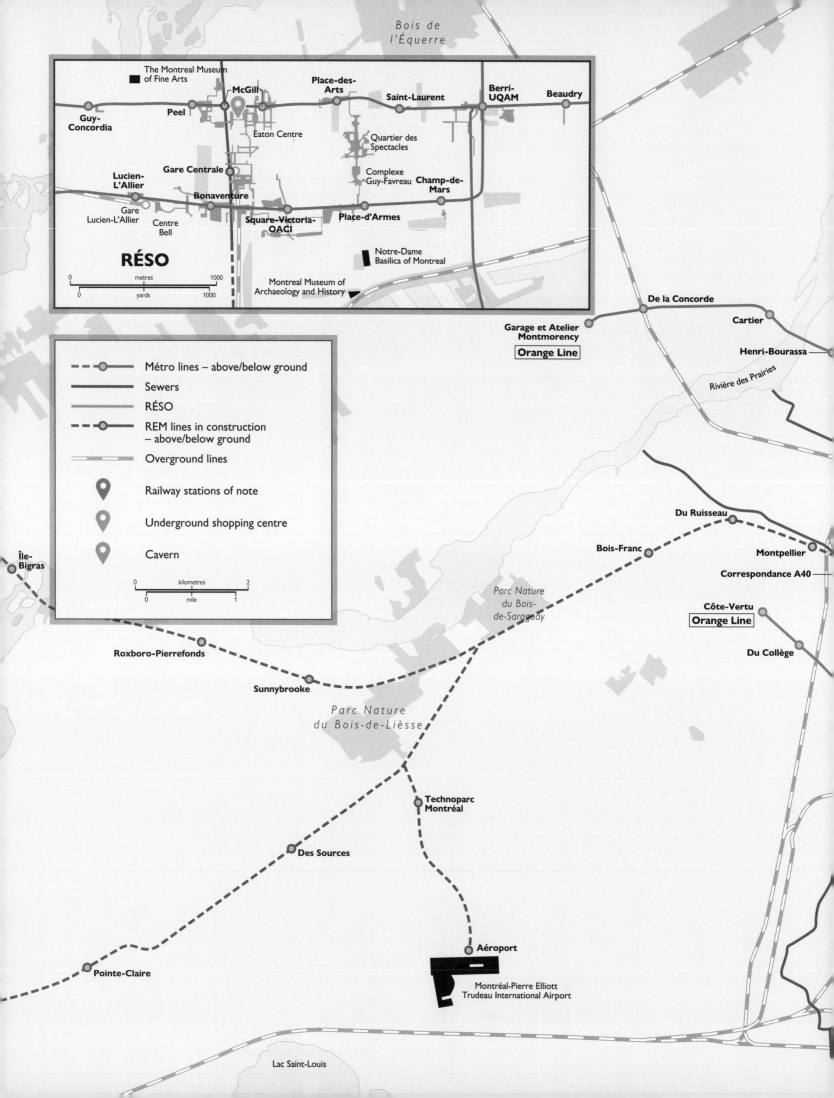

New York City

The global capital

The largest agglomeration in North America, New York City on the Eastern Seaboard of the United States, has a population of 8.6 million in its five main boroughs, and a further 20–23 million in the metropolitan hinterland. Its economy is so vast that if New York City was a separate country, it would be the world's twelfth wealthiest in terms of gross domestic product (GDP). While no city in North America has the span of historical habitation known to some of the European cities in this book, this global capital can certainly rival its counterparts for underground diversity.

City history

The original native New Yorkers were Algonquin hunter-gatherers who lived along the Hudson and Delaware river banks. In 1624 the Dutch West India Company set up a trading post on what is now called Governors Island, situated 730m (2,395ft) south of Manhattan in the Upper New York Bay of the Hudson River. Two years later the governor general Peter Minuit bought the neighbouring, and much larger, Manhattan Island from the natives, naming its first citadel New Amsterdam. Under British rule from 1664, the port at the tip of Manhattan was renamed after the Duke of York. It grew steadily to house 18,000 people by 1760, overtaking Boston. With a revised name of New York City, it become the temporary first capital of the embryonic United States for five months leading up to 1790.

Despite its leading role in battling colonial rule, the city retained strong ties with industrial Liverpool and Manchester in the United Kingdom, as a result of the import/export of cotton and the textile industry that drove the growth of all three. By 1840 the population had topped 200,000, making it one of the largest urban centres in the world. The five districts: Manhattan, The Bronx (the only part of the city not on an island), Brooklyn, Queens (both on Long Island) and Staten Island, were consolidated into a single entity in 1868 as the five boroughs of New York City.

The Statue of Liberty's torch-bearing arm was first shown at the Centennial International Exhibition of 1876, held in Philadelphia. Funds were raised for the rest of the statue and it was completed by the French and shipped across the Atlantic Ocean to be the full Lady Liberty in 1886. The statute stands on an old fort on Liberty Island, which is rumoured to house all kinds of secret passages, however a former worker disclosed that the only off-limit areas below ground were a salt store and an emergency walkway called Sally Port, that allowed visitors to exit quickly.

New York steam

One of the most clichéd images of New York is steam escaping from the streets. Much of this comes from the city's archaic, but effective, steam-generating and distribution network. It originated in 1882, when the New York Steam Company began sending high-powered, red-hot vapours around lower Manhattan in a subterranean network of pipes and tubes. Incredibly to outsiders, the system is still running, with 170km (106 miles) of active pipes pumping steam to 1,700 homes and businesses. It is not just used for heating either, but also for cleaning, and even cooling!

Early railroad tunnels

A tunnel runs for about 770m (2,525ft) underneath Brooklyn's Atlantic Street – between Boerum Place and Columbia Street. Named the Cobble Hill Tunnel, it is listed by Guinness World Records as North America's oldest underground rail structure. It was originally dug in 1845 by the Brooklyn and Jamaica Railroad (later the Long Island Railroad) as a 5m-deep (16-ft) cutting to get trains closer to the South Ferry landing stage on the East River. The trench was covered by 1850, but closed off in 1861. Police smashed their way in fifty-five years later, amid suspicions that it was being used as a bomb factory by German sympathizers during the First World War. When nothing was found the tunnel was sealed again, and despite various attempts to relocate it, the tunnel was not properly rediscovered until a twenty-year-old called Robert Diamond broke through a retaining wall in 1980 (with the help of a utility company) and started running tours of the abandoned structure.

Beginning around 1925, various routes were studied to cross the East River in order to link Brooklyn to downtown Manhattan via road. A bridge concept was considered too imposing, so work started in 1940 on the 2.7-km (1.7-mile) Brooklyn–Battery Toll Tunnel. Renamed after former governor of New York Hugh L. Carey in 2012, the tunnel took a decade to build and was the second-longest tunnel in the world at the time of opening (after Liverpool's Queensway Tunnel). Four huge ventilation shafts – one each in Brooklyn and on Governors Island and two in Manhattan – circulate air every ninety seconds. The structure was inundated during Hurricane Sandy in 2012, but quickly drained and reopened.

New York's first working underground railway was always intended as a trial operation (initially sold as a passenger-carrying version of London's pneumatic post system). It was opened in 1870 by the system's designer Alfred Ely Beach. Running for almost 100m (330ft) below Broadway, the vehicle was propelled using an ingenious system of air pressure. The Beach Pneumatic Transit Company ran the operation for three years, blowing an eager public – 400,000 of them in its first year – back and forth along the tube from its 'station' below Beach's offices at Warren Street to the stub end at Murray Street. There was no exit at the 'terminal', the intention was simply to demonstrate the effectiveness of the technology. The idea was eventually to build an 8-km (5-mile) tunnel to Central Park, but the stock market crashed and enthusiasm waned. The tunnel has been periodically rediscovered – for example, in 1912, when subway constructors working for the Brooklyn-Manhattan Transit Corporation (BMT) broke in whilst building the Broadway Line.

It is doubtful that air-blown systems would have been sufficiently robust for a full-scale mass transit system, but the concept was due to have another lease of life later on. Using similar air-powered technology, but on a much smaller scale,

BELOW Artist's impression of the bold Beach Pneumatic Transit vehicle. Although never intended as the only solution to the city street-based traffic chaos, Alfred Ely Beach used this demonstrator line to prove it could have been part of the answer and it holds the crown as the first attempt at underground public transit in New York City.

New York City gained a pneumatic mail-delivery system from 1897. As in schemes in Chicago, London, Manchester, Paris and Philadelphia, 0.5-m wide (1.5-ft) tubes were laid about 1m (3ft) below the surface either side of Manhattan Island. Inside them, pods containing posted documents were pushed along rapidly by air. A crosstown route was added from the new General Post Office to Grand Central station and another under the East River to Brooklyn in 1902. A total 43km (27 miles) of these pneumatic tubes made up a network capable of carrying almost 100,000 letters a day between twenty-three of the city's post offices. Despite frequent blockages and high ground rents, not to mention the onset of improved telephone and surface mail services, the system – suspended several times – limped on into the early 1950s.

The area between West 72nd and West 125th Streets (now known as the Upper West Side and Morningside Heights) was mostly railway land, but there was always plenty of under-utilized space between the rails and the Hudson River. The original Riverside Park had been created around the turn of the twentieth century, but it was a little cut off from the residential areas. In the 1930s city planner Robert Moses developed a scheme to bury New York Central Railroad tracks and the highway in 5km (3 miles) of tunnels and expand the park space over the top for easier access to nearby residents. The freight trains stopped using the tunnels in 1980 and homeless people moved in. Graffiti artist Chris 'Freedom' Pape started spraying the tunnel walls in 1974 and his artworks became legendary. In 1991, Amtrak tore down the shantytowns and reopened the tunnels to the railways, but they kept the name Freedom Tunnels.

Mass public transport

Such a huge metropolis wouldn't function without an intricate network of public transport. One in three of America's mass transit users and two-thirds of the nation's rail users are in this city alone. Counting the number of stations, the New York City Subway is the world's largest metro system, serving 424 in total (281 underground), although it is no longer the busiest system, the top five now being in Asia. The deepest station on the network is 191st Street, 55m (180ft) below ground.

The subway was not the city's first foray into mass transit: in the mid-1800s, all rail companies wanted to build elevated lines. The first to gain consent was the West Side and Yonkers Patent Railway running from the southern end of Manhattan to Cortlandt Street. After a wobbly start – it took until 1870 for the first fare-paying passenger to ride it – by 1880 there were four long lines along the major avenues. At its peak, the city was peppered with elevated railways. While these lines enabled the city to spread up Manhattan Island, they soon reached their limitations.

Proposals for underground railways had been circulating for years. New York State Legislature authorized the

ABOVE The newest part of the New York City Subway opened in 2017 with three stations, 72nd Street, 86th Street (pictured) and 96th Street. It is the first section of what will form the long planned Second Avenue Line – currently served as the northern terminus of Q Line – when the next sections open it will take the letter T. Plans for it date back a century when it was conceived to replace the elevated railway above ground. With platforms 28m (92ft) below the surface, 86th Street is air-conditioned and adorned with artworks by artist Chuck Close which include portraits of Lou Reed, Philip Glass and Cindy Sherman.

Metropolitan Railway Company in 1864, but the Bill granting rights to build a railway under Broadway from Battery to 34th Street and on to Central Park was quashed by the Senate. Despite Alfred Ely Beach's best efforts, his concept of a pneumatic tube was never going to cope with thousands of users. Finally, thirty years on, the first subway line was approved in 1894.

Although it took another decade for the Interborough Rapid Transit Company (IRT) line to open, civil engineer William Barclay Parsons had made the wisest decision in mass transit history: the route was four tracked to allow express trains to whizz past local stops, thereby massively speeding up the journey from one end of the island to the other – a huge advantage over the slower elevated lines. In 1904 the first part of the IRT subway line opened between City Hall and Grand Central. The most beautiful ceramic work was undertaken at the City Hall station, designed by architectural firm Heins & LaFarge. The tiling left people gasping that their new transit system would be beautiful, but sadly this was the most ornate station and the others only had ceramic-tiled friezes with station names and emblems on, also designed by Heins & LaFarge.

The Second Avenue Line

The Hudson & Manhattan Railroad (H&M) tunnels from Newark, New Jersey, into downtown Manhattan island opened in 1908 and are branded today as the PATH lines. Two other major operators, the Brooklyn–Manhattan Transit Corporation (BMT) and the Independent Subway System (IND) built many new lines, too. The history is complex and covered well elsewhere (see bibliography). The IRT, IND and BMT were unified in the 1940s and most of the elevated railways were torn down soon after. However, suprisingly few new lines or stations were added until recently.

A long-wished-for extension to a gap in the subway system between Harlem and Lower

ABOVE City Hall was originally perceived as key station of the early NYC Subway known of then as the IRT (Interborough Rapid Transit). It was the southernmost terminal of their Lexington Avenue line and the only one of its time to be built with curved platforms as it allowed trains to turn back to Brooklyn Bridge station without reversing. The decor was lavish: the ticket hall mezzanine and platforms were covered in ceramic tiles, and brass chandeliers held the lighting. It was closed in 1945 when the line was extended and the loop became redundant – though is still used for turning Line 6 trains.

Metres

0 — Pneumatic mail tubes
   — Sewers
   — Cobble Hill Tunnel
     Beach's pneumatic transit tunnel
-10 — Crown Finish Caves/
     New York Public Library storage

-20 — Times Square–42nd Street Station (IRT
     42nd Street Shuttle, BMT Broadway,
     IRT Seventh Avenue, IRT Flushing Lines)

   — Wall Street cylindrical vault

-30

-40

-50 — Croton Filtration Plant

   — 191st Street Station (IRT Seventh
     Avenue Line; deepest Subway station)

-60

-70

-80

-90

-100

-150 — New York City Water Tunnel No. 3

Times Square–42nd Street Station

Not surprisingly given its central location, this is the busiest station on the NYC Subway. This vast underground complex hosts almost 70 million passenger journeys a year. It is an interchange between four of the main trunk routes (and terminus of a shuttle) so therefore serves twelve 'lines'.

The complex essentially consists of five sets of platforms, four are shown in this simplified artist's illustration. At the end of the shuttle there is also a retractable bridge that allows passengers to cross over track four. Owing to its complexity and the cost of improvements the station complex has been the site of numerous renovations, the most recent of which will allow it to be fully accessible. The station is officially designated as: Times Square–42nd Street/Port Authority Bus Terminal station.

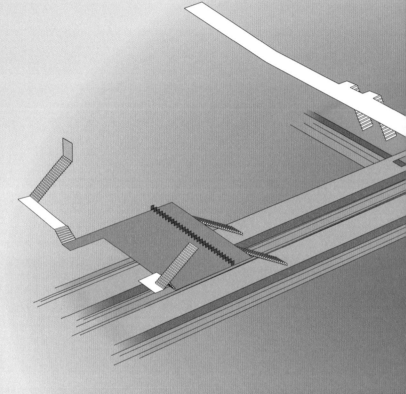

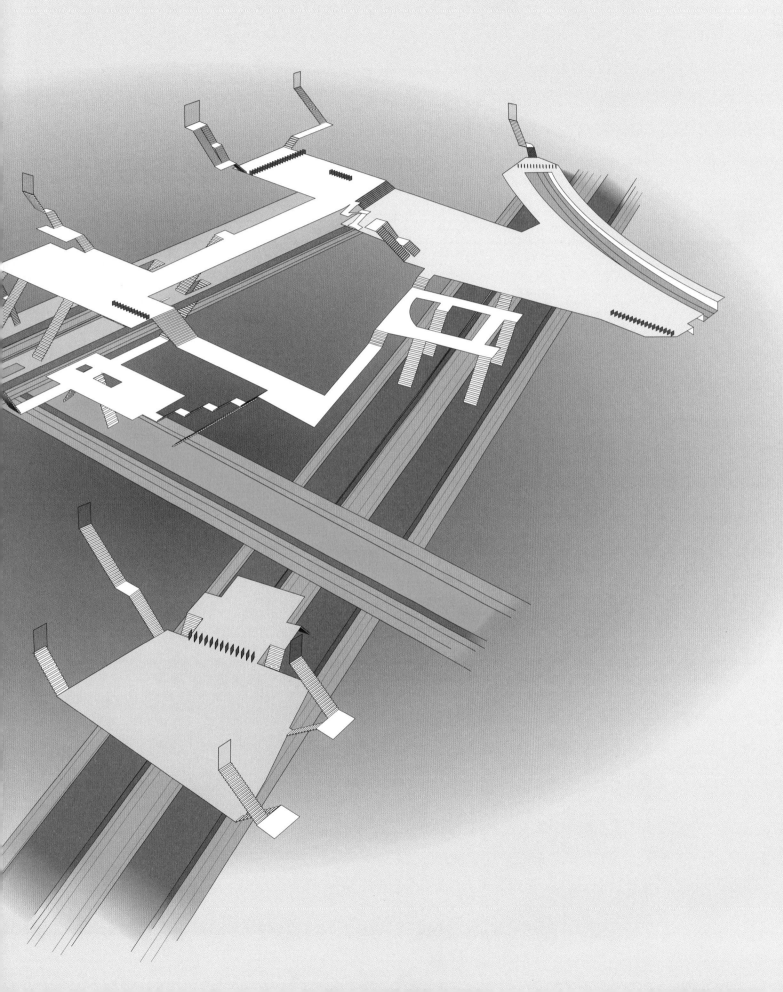

Manhattan had been dubbed the 'Second Avenue Line' owing to its trajectory, mostly beneath this East Side boulevard. Proposed as far back as 1919 and started in 1972, work stopped three years later because of lack of finance. Construction finally got going at the north end of the island in 2007. The first section of the new line opened a decade later, thanks to the utilization of an old section of rarely used tunnel under Central Park. The tunnel had formed part of a link between Queens and Manhattan that was constructed in the 1970s between Lexington Avenue and 63rd Street and Seventh Avenue and 57th Street. It was never brought into regular service, although it appeared briefly on maps in the 1990s. As the first new section of the Second Ave Line – now called the Second Avenue Subway – neared completion, a spur connected the unused tunnels to the new ones so that trains could run from the three new stops (at 96th, 86th and 72nd) onto the Q Line heading downtown. Opening in 2017, the three new stations (currently operated by the Q Line) will be joined by three more to 125th in the 2020s, when the line will assume the letter 'T' – though completion of the full southern section is still unfunded and many years off.

New York's ghosts

Owing to the expense required to lengthen platforms, trains stopped serving City Hall in 1945 and it remains one of several ghost stations beneath the city. Myrtle Avenue, on the Fourth Avenue BMT Line between the Manhattan Bridge and DeKalb Avenue in Brooklyn, was closed in 1956. On the walls opposite its abandoned platform, are more than 200 painted panels created by artist Bill Brand in 1980. As the train passes through the station, the panels work as a stop-motion animation to give a twenty-second 'movie' Brand called *Masstransiscope*.

East 18th Street station, once part of the first line opened in 1904, closed during a platform extension programme in 1948. A station at 91st Street was another casualty of platform extensions (closed in 1959), as was Worth Street station, which closed in 1962. Nevins Street station lower level was created to provide interchange for Brooklyn-bound locals, but no track was ever laid. More recently, the J/M/Z platform at Canal Street was closed in 2004.

Under the Waldorf-Astoria Hotel, built over the New York Central Railroad tracks going into Grand Central, is a special station that allowed wealthy guests with their own private rail cars to disembark discreetly into the hotel. From 'Track 61' a private freight elevator lifts the very richest of passengers directly into an exclusive hotel lobby. President F. D. Roosevelt, was among the frequent users and Andy Warhol held a party on the underground platform in 1965. A goods lift entrance down to long-closed platforms can still be found on East 49th Street.

In 1929 the South Fourth Street station was proposed on a line that would have linked Manhattan Island to Williamsburg via a tunnel beneath East River. Work was halted by the Great Depression and the Second World War, but in 2010 its never-used shell was taken over by street artists Workhorse and PAC who had created an underground street art expo called 'The Underbelly Project'. For 18 months more than 100 street artists from around the world transformed the space into a secret gallery – an event that is now the subject of a film.

Quirky underground spaces

New York City is littered with subterranean oddities that include the workings of a huge swimming pool at McCarren Park in Greenpoint, Brooklyn. Dating back to 1936, it was one of a dozen constructed by President Roosevelt's Works Progress Administration set up to keep people employed during the Great Depression.

Some 25m (82ft) below Wall Street is a two-storey cylindrical vault that rotates! It is said to contain up to 7,000 gold ingots in over 100 mini-vaults and is thought to be the biggest collection of gold in human history. The New York Federal Reserve Bank, which secures the bullion for financial institutions across the world, claims that it holds around 5 per cent of all of the gold ever mined and offers tours for anyone who can register in advance.

On Manhattan's Upper West Side, one of the oddest tunnels in New York City is 6½ Avenue. Running north–south between West 57th and West 51st street, this 400m (1,300ft) corridor is a secret passageway that's not shown in many tourist guides. The only street to incorporate a fraction in its name, 6½ Avenue is one of hundreds of Privately Owned Public Spaces (POPS) that originated in the 1960s. The POPS were created by New York City council to encourage property developers to create public spaces inside new buildings. Although 6½ Avenue is not strictly underground it feels like it is once inside.

Another odd tunnel runs beneath what was once a dangerous alley: Doyers Street in between Mott and Pell Streets in Chinatown. The tight-knit historic district was renowned for its interweaving passages laden with illicit gambling dens. The tunnel was supposedly excavated in the early 1900s to provide safety between there and nearby Chatham Square.

St Patricks Old Cathedral in Soho (built 1809–15) houses around thirty-five crypts and a handful of clerical vaults beneath its basilica. The catacombs are not generally open to visitors, but they are one of the last sites on Manhattan island that are still permitted to hold the remains of the dead. Other churches – for example, in Little Italy, have crypts too.

Food stories

The Meat-packing District didn't get its name by chance. During the nineteenth century, the western side of Manhattan close to 35th Street was peppered with slaughterhouses. Live cattle ferried over the Hudson River from New Jersey had to be herded along 12th Avenue, now West Side Highway, to be killed. The livestock proved too much for growing New York traffic, which even by the 1920s and 1930s was chock-full of private vehicles. So the story goes, a 60-m (195-ft) tunnel was made beneath 12th Avenue, just for the cows. However, anything left of the 1932 West Side Cow Pass would have been lost forever under the enormous footprint of the Javits Convention Center, built in the late 1970s.

In 1866, the owners of a Brooklyn brewery on Franklin Avenue in the Crown Heights area built an underground ice-storage holder a short distance away and connected it to the brewing area via a short 9m-deep (29.5-ft) brick-lined tunnel. The Nassau Brewing Company closed during the Prohibition era in 1916, but its Rundbogenstil (revivalist style) buildings were added to the National Register of Historic Places in 2014. The largering tunnel is now used to age 12,000kg (26,500lb) of cheese by a company called Crown Finish Caves.

BELOW The Crown Finish Caves are a subterranean cheese-aging facility in Crown Heights, Brooklyn. The former Nassau Brewery tunnels were perfect for adaptation to hold almost thirty thousand cheeses at any one time.

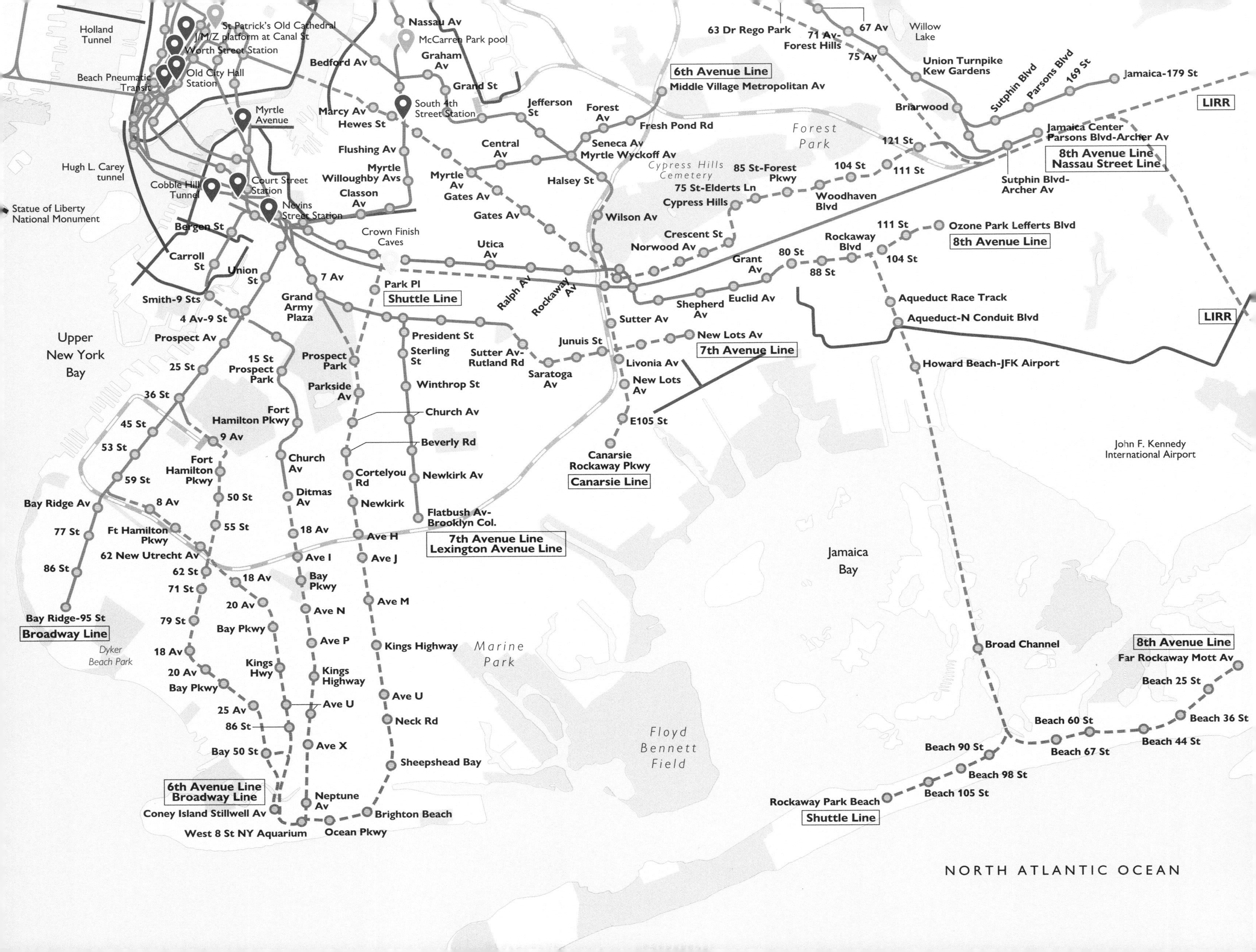

Holland Tunnel
St Patrick's Old Cathedral
J/M/Z platform at Canal St
Worth Street Station
Old City Hall Station
Beach Pneumatic Transit
Myrtle Avenue
Hugh L. Carey tunnel
Cobble Hill Tunnel
Court Street Station
Nevins Street Station
Statue of Liberty National Monument
Bergen St
Carroll St
Union St
Smith-9 Sts
4 Av-9 St
Upper New York Bay
Prospect Av
25 St
36 St
45 St
53 St
59 St
Bay Ridge Av
77 St
86 St
Bay Ridge-95 St
Broadway Line
Dyker Beach Park
Nassau Av
McCarren Park pool
Graham Av
Grand St
Bedford Av
Marcy Av
Hewes St
South 4th Street Station
Flushing Av
Myrtle Willoughby Avs
Classon Av
Crown Finish Caves
7 Av
Grand Army Plaza
Park Pl
Shuttle Line
15 St Prospect Park
Prospect Park
Parkside Av
Fort Hamilton Pkwy
9 Av
Fort Hamilton Pkwy
50 St
55 St
8 Av
Ft Hamilton Pkwy
62 New Utrecht Av
62 St
71 St
79 St
18 Av
20 Av
Bay Pkwy
25 Av
86 St
Bay 50 St
6th Avenue Line
Broadway Line
Coney Island Stillwell Av
West 8 St NY Aquarium
Church Av
Ditmas Av
18 Av
Ave I
Bay Pkwy
Ave N
Ave P
Kings Highway
Ave U
Ave X
Neptune Av
Ocean Pkwy
Kings Hwy
Brighton Beach
Sheepshead Bay
Neck Rd
Ave U
Kings Highway
Ave M
Ave J
Ave H
Newkirk
Cortelyou Rd
Church Av
Beverly Rd
Newkirk Av
Winthrop St
Sterling St
President St
Flatbush Av-Brooklyn Col.
7th Avenue Line
Lexington Avenue Line
Marine Park
Floyd Bennett Field
Jamaica Bay
Jefferson St
Central Av
Myrtle Av
Gates Av
Gates Av
Utica Av
Ralph Av
Rockaway Av
Sutter Av-Rutland Rd
Saratoga Av
Junius St
Halsey St
Forest Av
Seneca Av
Myrtle Wyckoff Av
Fresh Pond Rd
6th Avenue Line
Middle Village Metropolitan Av
Cypress Hills Cemetery
Wilson Av
Sutter Av
Livonia Av
New Lots Av
E105 St
Canarsie Rockaway Pkwy
Canarsie Line
New Lots Av
7th Avenue Line
Norwood Av
Crescent St
Cypress Hills
75 St-Elderts Ln
85 St-Forest Pkwy
Woodhaven Blvd
104 St
111 St
121 St
Forest Park
Shepherd Av
Euclid Av
Grant Av
80 St
88 St
Rockaway Blvd
104 St
111 St
Ozone Park Lefferts Blvd
8th Avenue Line
Aqueduct Race Track
Aqueduct-N Conduit Blvd
Howard Beach-JFK Airport
63 Dr Rego Park
71 Av-Forest Hills
67 Av
75 Av
Willow Lake
Union Turnpike Kew Gardens
Briarwood
Sutphin Blvd
Parsons Blvd
169 St
Jamaica-179 St
LIRR
Jamaica Center Parsons Blvd-Archer Av
8th Avenue Line
Nassau Street Line
Sutphin Blvd-Archer Av
LIRR
John F. Kennedy International Airport
Broad Channel
8th Avenue Line
Far Rockaway Mott Av
Beach 25 St
Beach 36 St
Beach 44 St
Beach 60 St
Beach 67 St
Beach 90 St
Beach 98 St
Beach 105 St
Rockaway Park Beach
Shuttle Line
NORTH ATLANTIC OCEAN

Boston

Tea Party was just the start

Situated by a large natural harbour on the North Atlantic coast of New England, and with four million people living in the metropolitan area, Boston is one of America's oldest cities. Founded in 1630 by English colonists, this was the site of the eponymous Tea Party that triggered the American Revolutionary War (1775–1783) during which the British were expelled from the city. Boston rapidly became the hub for New England's expanding roads and railways, a centre for the war against slavery and a mecca for intellectual, educational and cultural life. At one point Boston was as important as New York City for finance and the metropolitan area is still home to some of the country's greatest universities, including Harvard and Massachusetts Institute of Technology.

Subterranean curiosities

The tip of the Shawmut Peninsular on which the city is sited is known as North End. Dating back to the mid-1600s, it's the city's oldest residential area, and is rumoured to be riddled with secret tunnels. Certainly they were known about in the 1740s when a so-called Captain Gruchy smuggled four wooden statues ashore. Long closed off and sealed up, several tunnels are said to exist in cellars and were supposedly used by nineteenth-century 'rum runners' and bootleggers during the Prohibition era (1920–1933). One bar claims to have a tunnel running straight from Southie to the old shore.

Almost nothing remains of these ancient places today, owing to the extensive land reclamation and building that has taken place over the centuries, so urban archaeologists are always on the lookout for more to be unearthed.

Once dubbed the 'headquarters for the musical and artistic world of cultured Boston', the 1896 Steinert Hall was built entirely below surface level. The entrance beside the Steinway store on 'piano row' at 162 Boylston Street, took visitors four floors down into an auditorium deliberately sunk to cushion it from noisy streets. Although the curtain went down for the last time in 1942, there are plans to renovate the venue and bring it back to its old theatrical life.

Another feature of the old city that has been almost entirely obliterated by modern construction is the original brick sewer network: one running from Moon Island to the Dorchester area can still be located. Old streams have also been lost: Quincy's Town Brook, for example, makes a brief appearance at Star Market in the Quincy Center area, but the rest is culverted. Another old stream now covered over, Stony Brook can still be seen on the surface off Beaver Street, and is the run-off from Turtle Pond (a.k.a. Muddy Pond) – though it has never been full of turtles (or mud apparently) – which drains a water-logged and marshy area.

Early rail and road networks

By the late 1800s, electrification of Boston's tramways and new elevated railways provided the city with a sizeable public transport network. So many trams were passing through the central area, however, that it became necessary to put some of them underground. This led to the construction of

a tunnel beneath Tremont Street, running from Haymarket Square to Boylston at the edge of Boston Common. At Boylston, the service split to rise in two different directions: one towards Arlington and Back Bay and the other at Pleasant Street, where trams then split in two directions on the surface. Opened in 1897, the Tremont Street tram tunnel is effectively the oldest subway route still running outside Europe and the third oldest in the world to house electric traction propelled vehicles.

From the outset, the development of the Boston subway was a complex affair that created a large underground space at Park Street, which included a turning circle, and a peculiar platform arrangement at Schollay Square – both of which were much tampered with over the ensuing decades, resulting in the closure of the Pleasant Street spur in 1961. This left a sizeable section of abandoned tunnel, which is still in existence and fascinates Bostonians to the point that it has been considered for various reopening schemes. Another abandoned passage exists near the Boylston Street station and all five original Tremont Street subway stations have been altered, moved, resited or renamed. The main artery of this historic tunnel carries railed vehicles to this day, however, and now serves as the main downtown section of the Massachusetts Bay Transportation Authority's (MBTA) Green Line.

Three additional underground routes opened in the early 1900s: the East Boston Tunnel (1904; later the Blue Line), which burrowed under Boston Harbor; the Washington Street Tunnel, part of the Main Line Elevated (1908; later absorbed into the Orange Line); and

ABOVE The 'Public Garden Incline' of the Tremont Street Subway, shown here in 1904, was relocated a decade later.

BELOW Despite its awkward-looking trajectory beneath the Boston streets, this 1895 plan for the Tremont Street Subway would prove to be the first regular underground service of its kind in the USA. Much of it is still in use today, forming the crucial downtown section, of the now much longer Green Line.

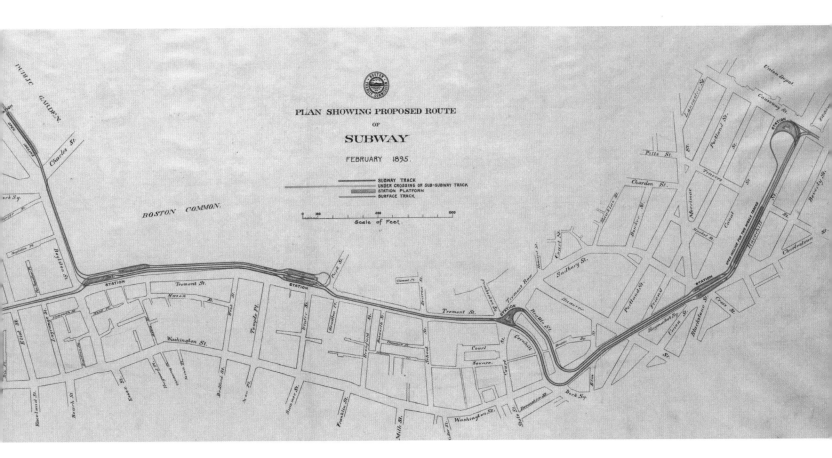

Metres

- 0
- Sumner Tunnel/Callahan Tunnel
- Stony Brook
- -10
- Steinert Hall
- Aquarium Station (Blue Line; deepest Subway station)
- -20
- Ted Williams Tunnel
- -30
- Central Artery
- -40
- Moon Island–Dorchester brick sewer
- -50
- -60
- -70
- -80
- -90
- -100

Ted Williams Tunnel and Steinart Hall

Boston prides itself on being a city of innovation and was once the largest urban area in what was then called British America, until it was overtaken by Philadelphia and New York City in the mid 1700s. It was home to the first public school, first public park and first subway in the United States. The city also claims to have the oldest continuously working streetcar system in the world. Despite this early start, Boston (like many other North American cities) went for Elevated Railways to solve the problem of getting commuters in and out of its central area. The Tremont Street Subway, along with the Cambridge and the Dorchester tunnels, now part of the Red Line, became incorporated within the wider system of rapid transit which encompassed elevated, street running and subway lines. This willingness to go below ground to solve transportation issues has been mirrored in a dozen tunnels for other traffic, especially road and bus.

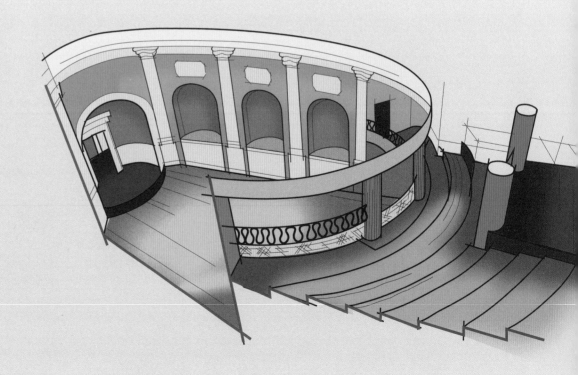

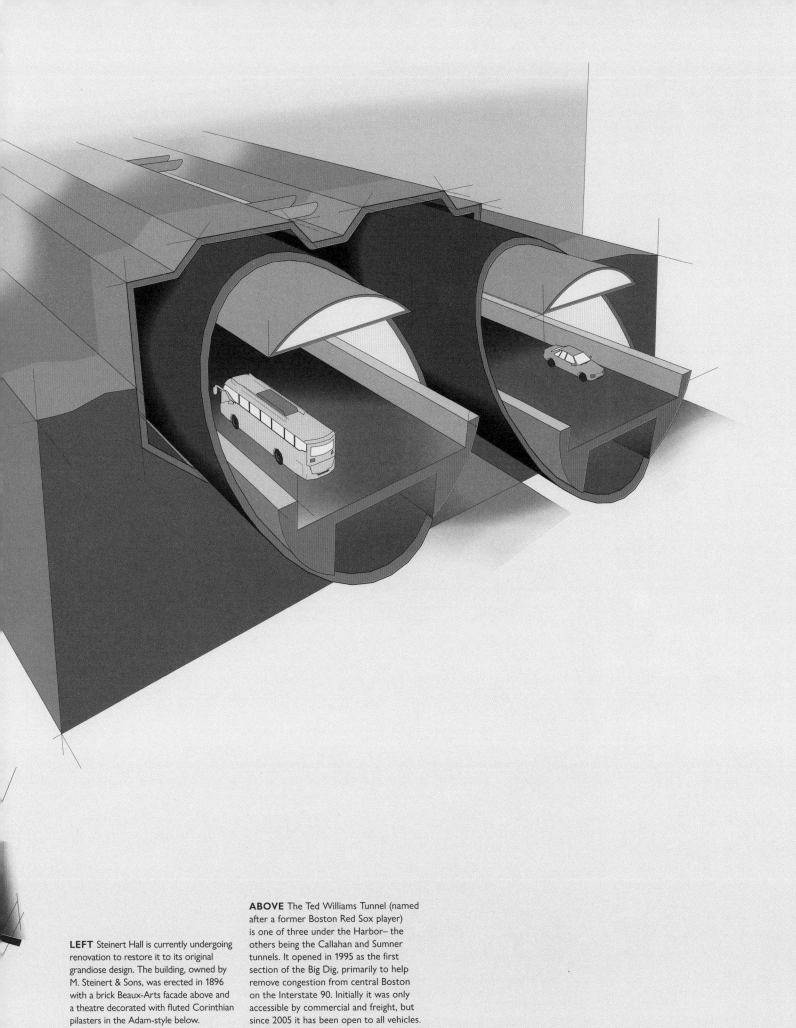

LEFT Steinert Hall is currently undergoing renovation to restore it to its original grandiose design. The building, owned by M. Steinert & Sons, was erected in 1896 with a brick Beaux-Arts facade above and a theatre decorated with fluted Corinthian pilasters in the Adam-style below.

ABOVE The Ted Williams Tunnel (named after a former Boston Red Sox player) is one of three under the Harbor– the others being the Callahan and Sumner tunnels. It opened in 1995 as the first section of the Big Dig, primarily to help remove congestion from central Boston on the Interstate 90. Initially it was only accessible by commercial and freight, but since 2005 it has been open to all vehicles.

the Cambridge–Dorchester Subway (1912; later used by the Red Line). While most of the original tunnelled stretches are still in use, there are several abandoned sections, such as the alternate platforms below the current Blue Line at Haymarket station.

In 1934 a new 1.7-km (1-mile) road tunnel was opened under Boston Harbor, between East Boston and North End. This undersea highway, known as the Sumner Tunnel, had one lane in each direction but rapidly became congested as private vehicle use increased across the United States. In 1961, therefore, the Callahan Tunnel was drilled parallel to the Sumner and now each functions as a one-way only highway.

Interstate I-90 runs across the USA, and the Boston to New York state line portion is known as The Massachusetts Turnpike, a 222-km (138-mile) east–west toll road. Beneath the road are tunnels connecting the booths at the Weston toll plaza, but they are only accessible by the staff.

The Big Dig

By far the biggest underground build in New England was the immense Central Artery/Tunnel Project (CA/T), known unofficially as the Big Dig. Conceived in the early 1980s, the idea was to extend Interstate 90 under Boston Harbor to Logan Airport and bury the elevated sections of Interstate 93 that ran through Boston's busy downtown area. Construction began in 1991 and took sixteen years to complete. In the process, it created the Rose Kennedy Greenway – a 2.4-km (1.5-mile) park that covers the tunnel, taking over the space once occupied by the highway. At a final cost of $14.6 billion, rising to $24 billon by the time the debt interest is paid off, it was the most expensive highway project in US history!

In conjunction with the Big Dig, a new tunnel was created for the city's Silver Line bus rapid transit system (BRT). Called the Ted Williams Tunnel, it is used by buses running from underneath South Station to Logan Airport. Another proposed tunnelled section for the BRT network would have consisted of an underground busway running from Washington Street to South Station and Boylston on the Green Line, but as the price tag rose to over $2 billion, the project otherwise known as the Little Dig is currently on hold.

ABOVE Boston operates a system of express buses, which run mostly in reserved lanes. The six routes of the Silver Line enter the downtown area in a specifically built bus tunnel which originally terminated at South Station, seen here. The route was recently extended to Chelsea via East Boston, partly in tunnel.

RIGHT The two photos were taken from similar vantage points before and after the Big Dig. They show just how much the road network impacted on Boston's downtown, and also how much the area benefits from open spaces and parks now that the roads have been buried.

BOSTON

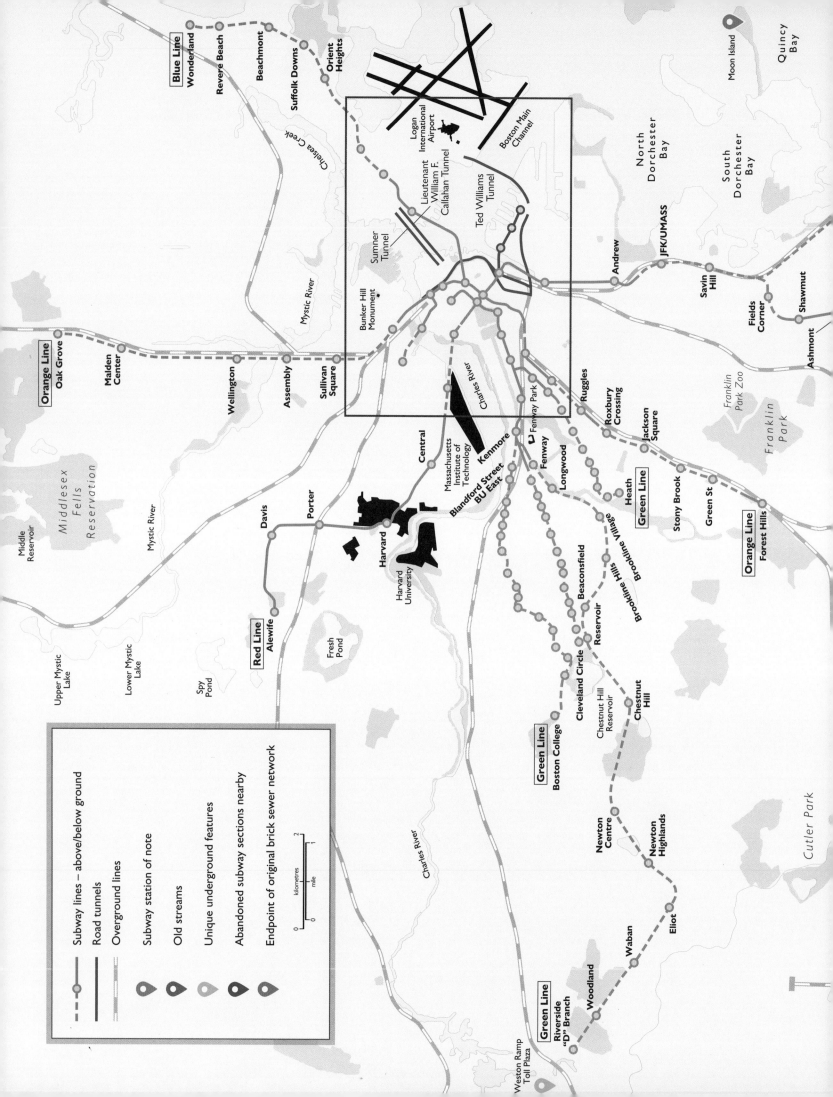

Buenos Aires

Persecution and perseverance

Argentina's capital region has a population of over twelve million, although Buenos Aires itself is an autonomous city within that area, with 2.9 million inhabitants. Founded by Spanish explorers in 1536 as Ciudad de Nuestra Señora Santa María del Buen Ayre, the area known as San Telmo, forms the most historic part of what is now Buenos Aires. Although this embryonic port lay abandoned after only a few years, it was resettled in 1580 as a trading post with the refined name, Puerto de Santa María de los Buenos Aires, subsequently becoming the hub of Spain's colonial expansion into the continent.

British forces attacked the city several times, capturing it in 1806, but it was soon liberated by Santiago de Liniers. It later became the capital of Argentina, which gained independence from Spain in 1816. At the end of the nineteenth century, the city experienced a cultural, economic and construction boom. Buenos Aires became the centre of the country's extensive railway network, cementing its position as the capital. The boom culminated in the opening of South America's first underground railway line in 1913, fuelling further expansion and a second boom in the early twentieth century.

Remnants of the past

Some of the city's earliest tunnels were excavated by Jesuit priests wary of persecution. They dug up to 2km (1.2 miles) of tunnel beneath the Manzana de las Luces (Block of Enlightenment). It has been postulated that these tunnels may have formed the first stage of an attempt to connect the area's major churches in order to provide the clergy and their congregations with an escape route in case of an assault on San Telmo. This network was not completed, however. The tunnels that do exist remained undiscovered for decades until renovation work started at a school within the Manzana de las Luces in 1912 and the floor gave way under the construction workers. It revealed an uncharted section of tunnel that archaeologists subsequently linked to the Jesuit priests.

During the seventeenth century, attempts were made to divert and cover over streams in San Telmo. These soon filled with discarded material and needed to be sunk deeper. Some were turned into modern sewers. Elsewhere, for example around Plaza Serrano, these works left large voids, which were subsequently built over. Some of these voids can be visited, and others have been turned into underground storage spaces. In one case, an extensive complex of tunnels and underground spaces in El Zanjón de Granados houses a museum and a banquet hall for corporate events.

Later in the twentieth century, the foundations of local landowner Jorge Eckstein's house became unstable and a patio caved in. On investigating, Eckstein discovered a huge maze of subterranean tunnels. These had been built later than the Manzana de las Luces tunnels, as part of the city's water treatment and flood prevention schemes, which started around 1780.

Buenos Aires Subte

Buenos Aires grew rapidly during the second half of the nineteenth century. At one point it laid

ABOVE The Subte is by far the oldest subway system in South America, having five lines before the end of the Second World War. This train heading to Plaza Italia was photographed in 1939 on the then relatively newly-opened Line D.

claim to the world's most extensive street-running tram system. Following electrification and intense competition between rival operators, some streets became so overcrowded with trams, however, that there were several plans to dig trenches along main arteries and bury the tram lines in tunnels. It wasn't until 1898, when work started on a new congress building, that the idea of better management of public transport finally gained momentum.

The former mayor of the city, Miguel Cané suggested in 1892 that an underground railway not dissimilar to the one that had been operating in London might be built, but others favoured an idea for surface-based suspended aerial trams. The subsurface option was finally chosen and, by 1913, the Compañía de Tramways Anglo Argentina (Anglo-Argentine Tramways Company) opened its first underground line. Meanwhile the Lacroze Hermanos company was lobbying to build a second line, which it began in 1927 and opened in 1930. Three more lines were then opened (C in 1930, D four years later and E in 1944) and the system now totals seven lines (six rapid transit and one light rail).

As with many older transit works, traffic patterns and operational needs change and Buenos Aires currently has four ghost stations – two of them on the first line (now Line A) and another two on Line E. On Line A, Alberti Norte and Pasco Sur both closed in 1953. Much of the former was destroyed when it was converted into an electricity substation, but the condition of the latter has been well maintained – when tourists visit, it gives the impression that the trains only recently stopped running. The ghost stations on Line E closed in 1966, when the trajectory of the line was altered away from Constitución station so that it could run closer to the heart of the city. While this helped to boost passenger numbers, it also left stops at Constitución and San José abandoned. There was a plan to resurrect them for a portion of the planned Line F, but the tunnels are now so old and curvaceous that it was considered unviable.

Buenos Aires Subte now has some extensive plans for future underground lines. Line H has recently been extended and the last section of improving an expanded major interchange at Retiro is finally underway.

Metres
0
  Manzana de las Luces tunnels
  El Zanjón de Granados
  Paseo del Bajo Road Corridor Project
-10

-20

-30

-40 Riachuelo plant wastewater tunnel

-50

-60

-70

-80

-90

-100

ABOVE Artist's impression of part of the huge basement of El Zanjon de Granados. Built over an ancient ravine in 1830 as a mansion for a wealthy family, the property was turned into a tenement by 1890 and the lower areas have recently been restored and opened to the public.

RIGHT The Paseo del Bajo Road Corridor Project, opened in 2019, is the culmination of a long-dreamt of attempt to take long-distance north-south commercial vehicles off the surface and turn formerly clogged roads into more accessible public open spaces. The new 7.1 km (4.4-mile) tunnel is reserved for lorries and buses, with only local traffic being allowed access to the surface roads and landscaped areas above.

El Zanjon de Granados and Paseo del Bajo Road Corridor Project

Given the city's pivotal location as the jumping off point for much of the European colonisation of South America, Buenos Aires naturally had a head start in subterranean development. The building of the railways and the Subte and their impressive speed of expansion (at least prior to the end of the 1940s, after which the pace slowed considerably) is an indication of the city's importance, wealth and population growth. Early excavations like the basement of El Zanjon de Granados have been lovingly restored in recent years and plans for new sub surface spaces are evolving.

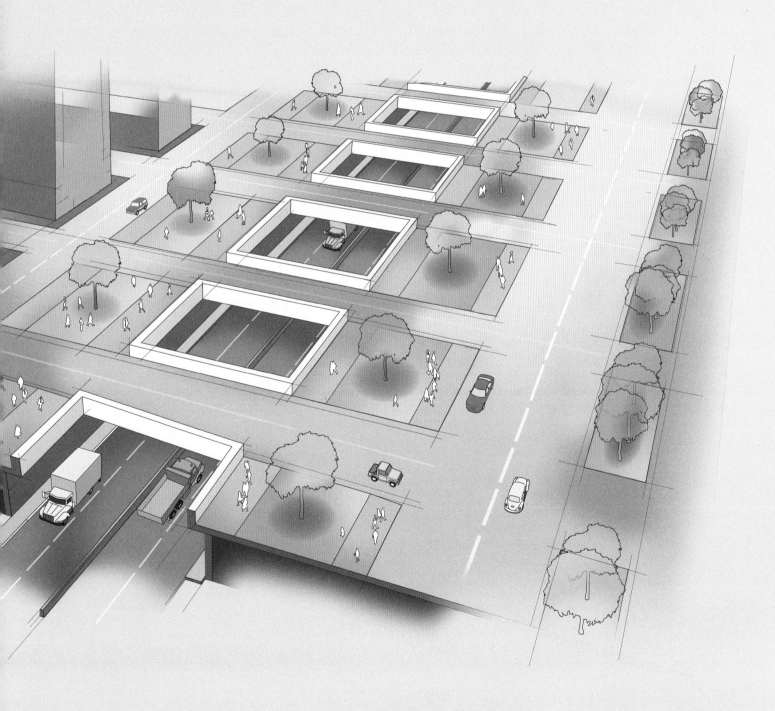

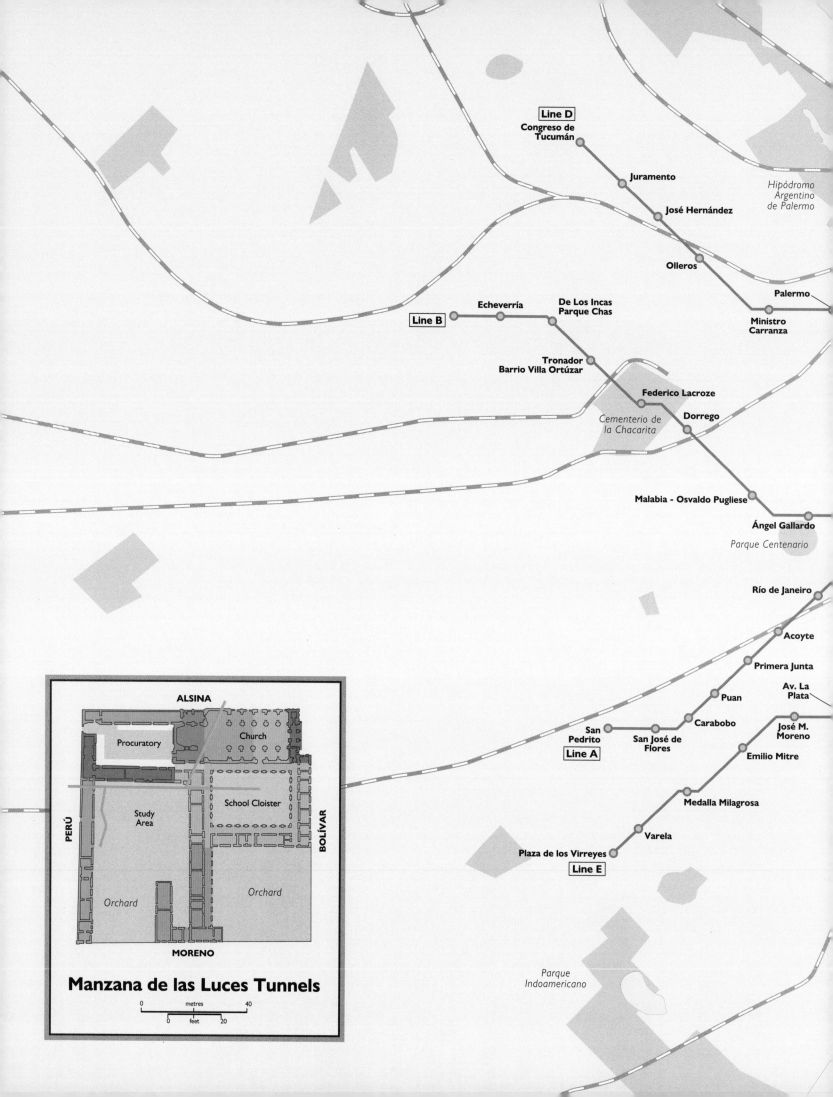

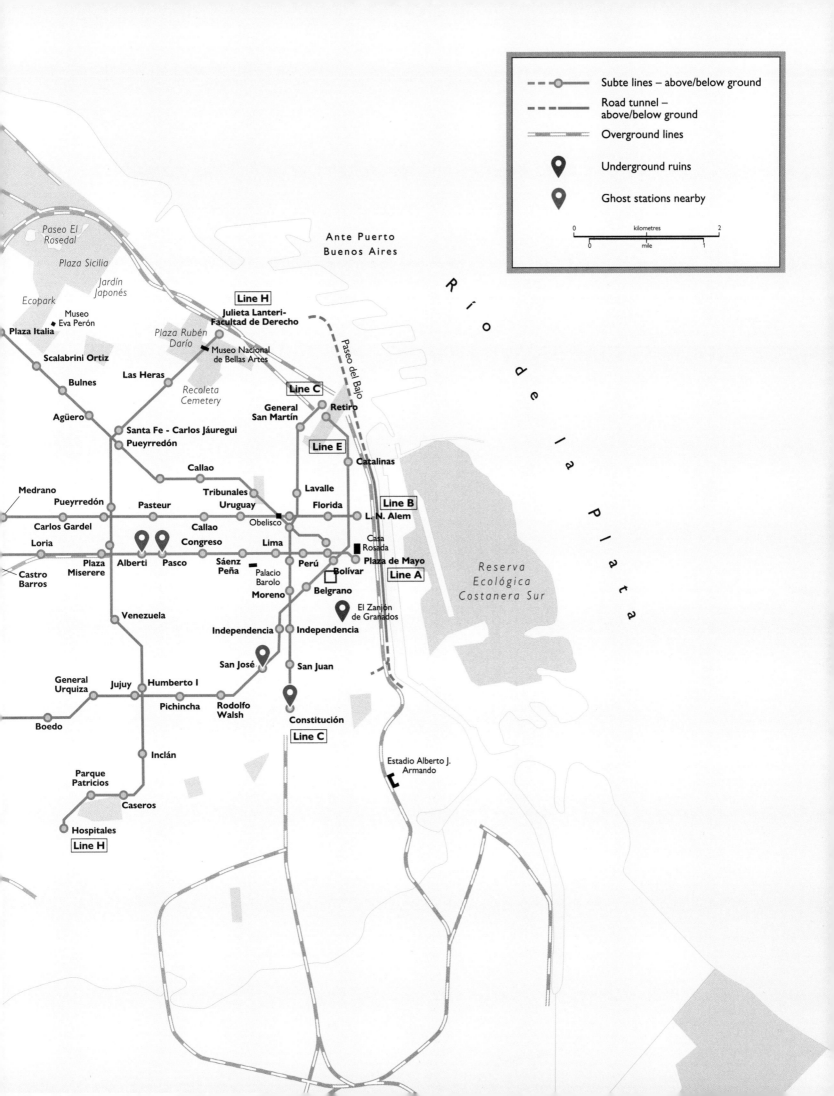

Europe

The pneumatic tube room in the The Central Telegraph Office,
St Martin's Le Grand, London, in the late 19th century.

Gibraltar

The riddled rock

The Gibraltar peninsular, commonly referred to by the British as 'the Rock', is a 426m-high (1,400-ft) limestone outcrop in a strategic position at the foot of Spain and overlooking the entrance to the Mediterranean Sea. Fought over for centuries and now home to 35,000 people, this landmass of just 6.8km² (2.6sq mi) is riddled with natural caves and around 55km (34 miles) of man-made tunnels. That's more than double the length of the roads on the Rock's surface.

Archaeological findings at Gothams Cave suggests that Neanderthals lived on Gibraltar 50,000 years ago. Early humans followed, and both the ancient Romans and ancient Greeks considered it to be one of the Pillars of Hercules flanking the Strait of Gibraltar – the other being on the North African mainland. The Moors built a castle here in the Middle Ages, and in the fifteenth century the Rock was won by the Spanish nobleman Juan Alonso de Guzmán in the Ninth Siege of Gibraltar.

Following the 1704 Capture of Gibraltar by Anglo-Dutch forces, excavations were made in the rock to provide defences. Passageways were carved out to connect gun emplacements, particularly on the northwestern flank, facing the Spanish mainland. Ceded to Britain in 1713, the Rock faced several attacks from the Spanish, and so yet more defences were dug during the eighteenth century. Construction happened in several phases, for example: the King's Lines, built during the 1620s on passages already started by Moorish troops; the Prince's Lines around a century later; and the Queen's Lines, begun in 1788. The lines were effectively trenches cut into the rock and protected by walls. Full-scale tunnels, sometimes called the Upper Galleries, were created by the British Army during the three years of The Great Siege of 1779–1783, when France and Spain joined forces in an attempt to oust the Brits, providing access to an otherwise innaccessable strategic outcrop at the northern end of the Rock. Gibraltar subsequently became a major base for the British Royal Navy, serving key roles in most of the biggest battles of the following centuries.

Another substantial phase of tunnelling took place between 1880 and 1915, providing access from Camp Bay to a quarry via the 1km-long (0.6-mile) long east–west Admiralty Tunnel (1883), and to two enlarged caves under Windmill Hill that were used to make an ammunition storage facility.

Clean water

During the nineteenth century, much excavation provided for desperately needed waterworks, including reservoirs that were made inside the Rock itself. Owing to a growth in population, the traditional method of collecting rainwater in a dishevelled array of cisterns and barrels led to regular outbreaks of diseases such as cholera and yellow fever. It is said that there was not a single clean water pipe in the town until as late as 1863.

Towards the end of the nineteenth century, the shortage of clean drinking water had become so acute that crude desalination of seawater was introduced. In the meantime, concrete underground water-catchment and storage tanks were created on Upper Rock and Sandy Bay. By 1903, 40,000m² (430,500ft²) of exposed rocky slopes had been covered with iron sheets to direct the precious rainwater into channels and stores. At its peak, more than 243,000m² (2,620,000ft²) of the terrain had been used for this purpose, but the system was discontinued in the 1990s and the rocks returned to their natural state. Today the drinking water is supplied by a dozen reservoirs buried inside

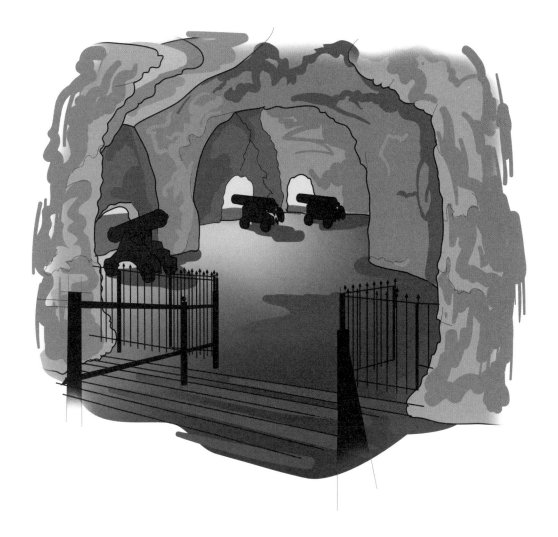

RIGHT Artist's impression of the Great Siege Tunnels. When captured Commander Duc de Crillon was shown the fortifications, he was so impressed that he conceded: 'These works are worthy of the Romans'. They are now a major tourist attraction to Gibraltar visitors.

the rock itself. The reservoirs were constructed at various times between the 1890s to the 1960s, and some are open to visitors; one even has an ornate balustrade walkway. A modern desalination plant now supplies 90 per cent of the potable water to the reservoirs and the remainder comes from rainwater-capture schemes.

A strategic base

Gibraltar's strategic location came into stark focus during the Spanish Civil War (1933–45) and the onset of the Second World War, during which it played a pivotal role. Almost the entire civilian population was evacuated and the Rock effectively turned into a fortress. It was from this time onwards that the lion's share of tunnelling was carried out. Waterworks, air-raid shelters and an underground hospital were built and the total length of tunnels increased from 8 to 11km (5 to 6.8 miles). From 1939 this soared to 40km (25 miles) when new space was created to house an entire garrison of troops, effectively creating an underground city.

Following military traditions, main thoroughfares were named after major British roads and included Fosse Way and Great North Road. AROW Street was named in honour of Lieutenant Colonel Arthur Robert Owen Williams of the Royal Engineers. Facilities included a bakery, a telephone exchange, a power station and enough space to store food and water for the 16,000 troops stationed within.

Many wartime constructions remained secret until long after hostilities had ceased. One of the most recent discoveries, made in 1997, was a so-called 'stay behind cave', which would have been staffed by six dedicated personnel who were to remain on the Rock in the event of a successful enemy invasion.

In response to the Cold War, a final phase of tunnelling happened between 1956 and 1968. Two more water reservoirs and a fuel-storage facility were created and the last, Molesend Way, was completed in 1967.

While many of the tunnels, including AROW Street, have been decommissioned and sealed off in recent years, many others – for example, the Upper Galleries and Second World War tunnels – are open to visitors. Two others, Dudley Ward Way (now closed) and Keightley Way (the last major one to be built) were assumed into the road network.

Madrid

Mazes and metros

The third-largest city in mainland Europe, the Spanish capital has a population of 3.3 million, with a further 3.5 million in the wider metropolitan area. It also has the continent's biggest underground metro system after London's.

Inhabited since prehistoric times, the seven hills on which the modern-day city of Madrid stands boasts Celtic, Roman and Visigoth remains. A riverside fortress from the mid-ninth century was built during Moorish rule and Christians took over in the eleventh century, leading to Madrid winning city rights in 1188. By the 1500s, there were already 30,000 inhabitants and the Spanish royal court moved to here from Toldeo. Railways arrived in 1851 and the population exploded, reaching half a million by the 1890s and topping a million by 1940.

Subterranean curiosities

From the tenth century onwards, the Moors built a canal network inside the hills, for water supply. Further tunnels were built in the hills to store food or weapons, as wine cellars, shelters or in some cases prisons. Some of these are known to have been deliberately carved out during the 1400s as secret escape routes from religious sites, palaces and military establishments. Up to 145km (90 miles) of subterranean tunnels have so far been discovered, but archaeologists suspect more will be found. Some had long air vents driven to the surface and topped by capirotes (little turrets similar to pointed hats of the same name), which can still be viewed in places such as Dehesa de la Villa and Fuente del Berro parks. Most of these tunnels are quite winding and decrepit, but a few, such as one built in 1809 between the Royal Palace (former site of the Habsburg Alcázar) and Casa de Campo, and 45m (148ft) long, was neatly finished with brickwork and well lit. It was made for King Joseph Bonaparte.

From the 1870s to the 1880s King Alfonso XII also utilized a maze of tunnels beneath the Royal Palace, as a means of escaping the grounds to go on nefarious nocturnal outings.

Clandestine activities

At Plaza de la Marina Española, a space beneath the Senate served as a shooting range during the nineteenth century, used by a nearby military barracks for training. The location, alongside a building used by the Spanish Inquisition until around 1820, was littered with underground dungeons. The space also proved helpful during the twentieth century, as a personal bunker for the Spanish dictator Francisco Franco in 1946, during the most delicate period of his reign.

In deep tunnels beneath the old building housing the SER network (originally Unión Radio studios) on the main boulevard Gran Vía, is a basement that housed a clandestine radio station run by the so-called Fifth Columnists during the Spanish Civil War. They used it to broadcast Republican intelligence information from the rearguard to help Franco's campaign.

On the ground floor of the Finance Ministry on Alcalá street, a strange door with metal bars over it

ABOVE Alto de Arenal station has one of the Metro's earliest carriages on display in an unusual manner, suspended above the running tracks. Visitors can see this 1928 car sitting proudly on its own piece of track as they ascend the escalators.

– accessed via the Pasaje de la Caja de Ahorros – has a ladder dropping down to a deep subterranean space where the famed socialist politician Julián Besteiro broadcast news of the opposition forces in the period leading up to the advancing armies in 1939.

City transport

With the population soaring, trams were brought to Madrid's streets in 1871. Horse-drawn at first, a steam tram appeared in 1879 and from 1899 trams were mostly converted to electric power. As often happens with such expansion, plans either to put the trams in tunnels or to build a rapid transit system below ground were proposed. A north–south project started in 1916. The first section of the new Metro system opened in 1919, serving eight stops between Sol and Cuartro Caminos – just under 4km (2.5 miles). The next part, between Sol and the main rail station at Atocha, arrived two years later. By 1936 there were three lines plus a branch (or shuttle), between Norte and Opera. These original lines were extended only a little every decade, until 1995–2007, when there was a huge expansion of eighty new stations. This included the MetroSur – a deep-level 41-km (25.5-mile) circular line serving five southern suburbs, which opened in 2003. The Madrid Metro now boasts 293km (182 miles) and 301 stations (the twelfth most extensive in the world). Owing to the Spanish financial crisis of 2009, further extensions are now on hold.

As in other cities, the network in Madrid has its fair share of oddities. In 1961 a new concept was proposed to take rapid transit further out from the centre of the city: the Suburbano. Only a short section of tunnel was built, which has now been assimilated into Madrid Metro's Line 10. There is only one closed station on the Madrid system: Chamberí. It was shut in 1966 owing to its proximity to Iglesia and Bilbao stations, but also because the cost and difficulty of extending its curved platforms proved prohibitive. Since 2006, it has been made safe and cleaned up – including restoration work to the old ceramic tile adverts – to host a museum, named Andén 0. It faithfully recreates the ambience that was designed by the original architect Antonio Palacios in 1919. There is an abandoned tunnel under Chamartín Railway Station, which was built, but not used, for Line 1. A station on the Line 7 extension that was prepared in 1999 would have been called Arroyo del Fresno, but was not finished or opened, as there were not enough residents and businesses at ground level to support it at the time.

Further transport links arrived in 2007, when Madrid brought modern trams to the streets with the opening of the first line of Metro Ligero, a light-rail system with long underground sections including 10km (6 miles) in tunnel. It now has four lines covering 36km (22 miles). Line ML-1 feels like an underground railway as the majority of stations are in tunnels.

Dormitory towns around Madrid started later than elsewhere. Known locally as 'bedroom communities', they suffered from poorly used local rail lines until a new

BELOW Chamberí station was opened in 1919 with the initial run of eight stops on Spain's first true Metro line. During the 1960s the whole line needed to be upgraded to take longer trains, but at Chamberí it was not physically possible to do this owing to its proximity to stations either side (Iglesia and Bilbao), so it had to be closed in 1966. The platforms were bricked up so that trains could pass through at normal line speed which preserved the interior. Despite being vandalized, restoration began in 2006, and two years later it re-opened as a museum to Metro history complete with heritage signage and advertising panels still in place.

ABOVE Redirecting parts of Madrid's M-30 motorway underground have allowed for much-needed road refurbishment and increased capacity, but also bring hopes of reducing surface pollution.

underground link was built between Atocha and Chamartín by Línea de Enlaces Ferroviarios, which was finished in 1967. It was rebranded as a Cercanías C-1 and C-2 by the national rail operator RENFE, having established this new name for its suburban and commuter services in the 1990s. A second underground link, also between Atocha and Chamartín, but with an intermediate station at Sol to connect with the busy Metro interchange, was opened in 2008 and is used by C-3 and C-4.

Having embraced the concept of high-speed rail, Spain aggressively developed a large network across the country. At Madrid, the lines terminated either side of the city, so a new 7.3-km (4.5-ft) tunnel connecting the high-speed lines entering the city from the north and the south is set to open soon. The tunnel runs at an average depth of 45m (148ft), burrowing beneath eight different Metro tunnels and two Cercanías tunnels between Atocha and Charmartín.

The orbital motorway around the capital, the 32.5-km (20-mile) Autopista de Circunvalación M-30 was started in the 1960s and required the Abroñigal River to be placed in a culvert to avoid it flooding onto the roadway. It took almost thirty years to complete the construction of this vital inner relief road, which needed to be upgraded by 2005. More than 10km (6 miles) of this huge new highway was placed underground, with one individual tunnelled section being over 6km (4 miles) long. It is now Spain's busiest road.

Liverpool

First rail tunnels

One of Britain's best-known provincial cities, England's northwestern port of Liverpool is also a sporting and cultural centre. Half a million people live in the city, with a further million in the wider metropolitan area.

The earliest known settlement was referred to as Liuerpul as far back as 1190 and the English King John apparently designed a plan of seven streets in 1207. Despite a royal charter, it took several hundred years for the fledgling port to rival the older port city of Chester. One of the changes to Liverpool's fortunes came when the first slave ship left for Africa in 1699 and trade with the West Indies grew. At the same time, Chester became inaccessible as the River Dee became increasingly clogged with silt. Liverpool's first wet dock opened in 1715 and import/export became the root of the city's wealth, upon which prominent businesses built lavish buildings in neoclassical style along the surrounding streets.

Making railway history

Most of the goods passsing through Liverpool's docks were bound for the burgeoning heart of the Industrial Revolution, just a few miles inland from Liverpool – the textile mill towns of south Lancashire, especially those around rapidly expanding Manchester.

The volume of freight passing back and forth at this time gave rise to Liverpool's pioneering role in railway history, as it is home to the world's first major underground rail infrastructure. The historic heart of the city sits on a sandstone hill, so when the first inter-urban railway line was built between Liverpool and Manchester in the late 1820s, it necessitated placing some tracks in deep cuttings and tunnels. The Liverpool & Manchester Railway (L&MR) built the first passenger station at Crown Street, close to Edge Hill to the east of the main city centre. Here, engineer George Stevenson devised the 262-m (860-ft) Cavendish cutting and a short piece of tunnel that were made for the station. But to move freight the extra distance down to the waterside, Stephenson designed the unprecedented 2-km (1.2-mile) Wapping Tunnel, which was bored between 1826 and 1829 under the city to South End Docks goods station (later renamed Park Lane station).

The line opened to passengers and freight in 1830, but the Wapping Tunnel was too steep for the crude locomotives of the day, so a steam-powered cable hauled trains from the docks up to Edge Hill, where they were hitched to a locomotive for the journey to Manchester.

Just six years later, a new 1-km (1.1-mile) tunnel was needed to get passenger trains from Edge Hill to a brand new station called Lime Street closer to the heart of the city. And in 1848, the 4.3-km (2.7-mile) Victoria-Waterloo tunnel was dug to get freight to the North End docks. This meant that, within a few short years, three tunnels fanned out from Edge Hill underneath Liverpool city centre, turning it into an important rail hub.

The docks themselves were crucial to Liverpool's existence and had such a long frontage that it was necessary to build a railway to connect them all together. Plans for a surface railway looked horribly complicated with the number of crossings needed to serve each dock, so an elevated line, the Liverpool Overhead was proposed as early as 1852, although work didn't start until 1889. Four years later 11km (6.8 miles) of elevated structures between Herculaneum and Alexandra Docks had been completed and track laid on top. The network benefitted from the delay, becoming the world's first electrically powered overhead rail system.

A short 800-m (2,600-ft) extension was opened in 1896, taking the railway away from the docks and through a tunnel to Dingle, where many dockers lived. The plan was to continue the tunnel further inland with more stations serving residential areas, but this did not materialize and Dingle remained an oddity: an underground station on an elevated line. The whole system closed in 1956.

The Edge Hill freight yard, like Liverpool Overhead, was closed in 1972, along with the original 1829 tunnels by Stephenson. The Crown Street portal is now landscaped over. The Wapping Tunnel has also been disused since 1972.

Towards a city network

Liverpool Docks are situated on the Mersey River estuary, which faces the Wirral peninsular. During the nineteenth century, the town of Birkenhead grew rapidly here, on the shore facing Liverpool, and was famously reached via the Mersey Ferry. In 1860, Birkenhead built the United Kingdom's first tramway, which improved transport in the town, but did not link it to Liverpool. Needing to cross 1km (0.6 miles) of water, the idea of a building a bridge seemed impossible so, in 1871, the Mersey Railway company gained permission to begin work on a tunnel beneath the river.

Following an immense construction effort, which included lining the tunnel with thirty-eight million bricks, the new railway serving four stations was finally ready in 1886. Two of the stations, Hamilton Square and James Street, are effectively the oldest deep-level rail stations in the world. The network was extended on the Birkenhead side and on to Liverpool Central in 1892, so that trains could run right through the tunnel from other routes. Despite huge steam-powered fans on either side of the river, the tunnel became heavily polluted by the locomotives' exhausts.

RIGHT Thomas Talbot Bury's 1833 watercolour of the Liverpool & Manchester Railway tunnel portals at Edge Hill, which were effectively the world's first ever rail infrastructure under the ground. Located in a cutting, the station is 23m (75.5ft) down. The right hand tunnel descended to a passenger terminus at Crown Street; the left hand one led to a short siding but was created merely for artistic merit. The middle tunnel led to Wapping Docks. The two chimneys (known as 'Pillars of Hercules') released smoke generated by the stationary steam engines that pulled the cable which helped to move trains back and forth to Crown Street.

Many passengers opted to use the Mersey Ferry as a result and, by 1900, Mersey Railway had gone bankrupt. The solution was electrification, which allowed the connection to reopen in 1903, after which train speeds increased and the system became popular again.

In the early 1970s a 'loop and link' project was built to integrate Merseyside's separate routes, the Northern and the Wirral Lines. The Loop was a new, single-track tunnel from the former Mersey Railway, now the Wirral Line, which would divert trains in a circuit below Moorfields, Lime Street and Liverpool Central then back to James Street, where it rejoined the original tunnel under the Mersey. The Link was a double-track tunnel joining the Hunts Cross branch of the Northern Line with the Southport branch heading north. The huge project was completed in 1977 and provides the city with one of Britain's only underground railways outside the capital. A short section of the Mersey Railway between James Street and Liverpool Central had to be closed for regular passenger use in the process.

Designed in the 1920s, the first road crossing beneath the Mersey, between Liverpool and Birkenhead, was the 3.2-km (2-mile) Queensway Tunnel, which finally opened in 1934. It became the longest road tunnel in the world. Entrance portals, lamps, ventilation towers and toll booths were all finished in the art deco style. A relief route planned in the 1960s – the 2.4-km (1.5-mile) dual Kingsway Tunnel – opened in 1971, between Liverpool and Wallasey. A short section of the Queensway tunnel, on the Birkenhead side, (Rendel Street branch) was closed in 1965.

As a safety precaution in case of fire or flooding, seven emergency refuges have been created underneath the road deck. Linked to each other via walkways and with exits at both ends, each is able to hold up to 180 people.

One man's curiosity

Liverpool landowner, tobacco merchant and philanthropist Joseph Williamson embarked on some tunnelling fun of his own. Between 1810 and 1840, Williamson built a curious series of subterranean passages and openings near to Edge Hill. Now known as the Williamson Tunnels, most of the excavations are lined with brick or

LIVERPOOL

ABOVE Artist's impression of the high vaulted ceiling of one of Williamson's Tunnels.

TOP LEFT The road tunnel under the Mersey was so complex it took nine years to achieve. This 1926 drawing from the *Illustrated London News* suggests a double-deck arrangement that was not how the structure was delivered.

BOTTOM LEFT When the Queensway Tunnel was finally ready for opening in July 1934 by King George IV and Queen Mary it was nicknamed the 'Eighth Wonder of the World' and 200,000 turned out to witness it.

stone. The exact purpose of these tunnels – beyond the sheer joy of Williamson's experimentations – was unclear until a cavernous banqueting hall was recently discovered. The work seems to have been carried out by Williamson's workers after they had created a series of brick arches on which to lay a formal garden for his house. But Williamson kept the staff on, creating vast caverns that reportedly included two large underground 'houses' connected to each other via a spiral staircase. After Williamson's death the spaces began to clog up with rubbish, waste water and even sewage, resulting in them needing to be backfilled with rubble and sealed. A society was founded in his name in 1989 and now maintains what is left. In the early 2000s, the organization opened a number of restored tunnels to the public.

Manchester

Pioneers and pipedreams

The capital of northern England and the world's first industrial city, Manchester can also lay claim to being the birthplace of modern computing, atomic theory, female suffrage, trade unions, the vegetarian movement and a plethora of underground spaces. It has a city population of 560,000, with an additional two million in the metropolitan area.

Founded by the ancient Romans as Mancunium, the city grew slowly through the Middle Ages, but as steam power and textile manufacture took hold in the late-eighteenth century, the city population expanded rapidly from a few thousand to half a million by 1900.

Manchester's canals

Britain's first major modern canal, the Bridgewater, opened in 1761 and ran from the suburb of Worsley to the inner-city district of Castlefield. Its success inspired a spate of waterway building dubbed Canal Mania. The creation of the canal basin at Castlefield required the Mersey River to be diverted into an underground culvert. The city became criss-crossed by canals and one plan in 1799 would have seen a 1-km (0.6-mile) long tunnel link the Irwell River with the Rochdale Canal, but a parliamentary Bill did not happen for thirty-six years. When it did come, the Manchester and Salford Junction Canal tunnel ran 11m (36ft) below Camp Street – with two loading/unloading docks under the Great Northern Railway Warehouse (still standing on Deansgate) – and beneath what later became Central Station and Lower Mosely Street to a connection with the Rochdale Canal. The Manchester and Salford Junction Canal was drained during the Second World War and functioned as a huge air-raid shelter big enough for hundreds of people to sleep. Owing to the city's pre-eminence in industrial growth, there are many other waterways buried beneath its streets: The River Tib was culverted in 1783. The 600-m (2,000-ft) long Dukes Tunnel was built in 1789 to transport coal brought along the River Medlock from the mines in Worsley to Bank Top. At Bank Top, the coal was raised from below ground via a deep shaft and transferred onto barges for transporting along the Rochdale Canal. Dukes Tunnel was abandoned after just twenty years when, in 1800, new, more direct canals were built to traverse the city centre. Several other tunnels were built for barges, including the Bengal Street tunnel to Ashton Basin in Ancoats, which opened in 1798.

Scuppered metro plans

Of all the cities in the United Kingdom that have tried the hardest – and failed the most – to build underground railways, it is Manchester. At least a dozen serious proposals emerged over the years, including one that would have been the first in the world. In 1830 a new inter-urban railway connecting Manchester with Liverpool was such a success that the initial station was deemed too far from the heart of the city and needed resiting. A line was quickly extended to a site near Manchester Cathedral at Hunts Bank, where

ABOVE The brick vaulted, gas-lit tunnel of the Manchester & Salford Junction Canal was almost 500m (0.3miles) long and opened in 1839. Running roughly east/west under present day Camp Street, it was partially shut in 1875 when the vast Central Station was constructed above ground. The truncated section finally closed in 1936, but the structure was drained and given a new lease of life as a huge air raid shelter for up to 1,350 people, seen here not long after the horrendous Manchester Blitz of December 1940. It is now a Grade II listed building.

Manchester Victoria station was opened in 1844. By 1847, the Manchester and Birmingham Railway was due to arrive at Store Street (later named Piccadilly), about 1.5km (0.9 miles) south of there, and so both companies saw merit in linking their distant termini. They planned a tunnel between the two, and this might have formed the start of an underground railway, had it materialized.

Additional underground rail and tram tunnel plans emerged in 1868, 1878, 1903, 1911 and 1914, along with four different schemes in the 1920s, four more in the 1930s, several in the 1950s and no less than seven in the 1960s. Finally, work actually got underway in the early 1970s: an underground mainline rail link dubbed Picc-Vic was to be built between Piccadilly and Victoria with five subterranean stations. Ground surveys were carried out, 3D models of stations and tunnels were built, architects' plans were issued, stations designed, bore holes dug – almost twenty of them down as far as 27m (89ft) – and space was left for future platforms under the new Arndale shopping mall. But while both Newcastle and Liverpool got their underground rail improvements built, Manchester's ambitious plans were scuppered during the 1974 oil crisis, when the government withdrew funding. The ambitious Picc-Vic project withered and died.

However, at least one railway in a tunnel did get built and survived – the 1.5-km (1-mile) Salford Underground Railway. Opened by the Lancashire and Yorkshire Railway, in 1898, it ran from the Manchester Ship Canal at Trafford Road (between docks 8 and 9) to a point close to the current Windsor Link. It closed in 1963. Originally it had a platform for passengers at the docks end (New Barns station) but no trains served it after 1901. Today, a few short sections of tracks are present in tunnels on the current Metrolink light-rail network that opened in 1992. Greater Manchester Mayor Andy Burnham resurrected the underground railway concept again in 2019, promising that any future Metrolink expansion in the city centre would need to be made in tunnels.

Hidden gems

Between Manchester Cathedral and the Irwell River, a series of embankments had existed since the early days of city settlement. The layout changed several times and the feature became known as the Cathedral Steps. Between 1833 and 1838 a new embankment was added when a row of brick arches was built to lift the surface to a higher level. Named the Victoria Arches, they contained public toilets, a wine cellar and small shops. By the end of the nineteenth century, wooden landing stages had been erected from the water's edge to allow small boats to dock and provide pleasure trips along the river towards the Manchester Ship Canal and the Pomona area. As the river became more polluted, the arches fell into disrepair and were bricked up, although they were repurposed as air-raid shelters for more than 1,600 people in 1939, in preparation for the coming war.

One of the longest, straightest roads in the city, Deansgate is home to many shops, including the fabled Kendal Milne department store, the owners of which themselves dug a tunnel between two parts of their premises in 1921. A 2.5-km (1.5-mile) tunnel, up to 20m (66ft) deep in places, is said to run below the street from the Cathedral via Knott Mill under Chester Road to Cornbrook. Neither the date nor the purpose of the tunnel has ever been firmly established.

For its first 138 years The Guardian newspaper was published in Manchester. It relocated to London in 1959 (dropping the name Manchester from its title) but not before a Cold War construction had also adopted the 'guardian' name. During the 1950s the Post Office hatched a plan to build three vital, nuclear-bombproof communications facilities in the key cities of London, Birmingham and Manchester. The Guardian Underground Telephone Exchange

BELOW Although the threat of nuclear war (for which the deep-level Guardian Telephone Exchange had been built) was fizzling out through the 1980s, it continued as a major national routing centre for BT until 1988 as this image from five years earlier shows. The tunnelled structure housed not only the telephone exchange equipment, but also sleeping quarters and a kitchen for staff, diesel-powered generators and even fake painted windows and a pool table. Although now defunct, the tunnels still house many phone cables.

ABOVE Standing on one of the three nearby bridges or on the opposite bank of the Irwell it is just possible to make out the bricked up entrances to what was once a multi-purpose series of Victorian arches, sometimes known as Cathedral Steps. Repurposed during the Second World War as air raid shelters, there is now talk of refurbishing them again as a tourist attraction.

(GUTE) a.k.a. Scheme 567 was excavated between 1953 and 1957 at 34m (112ft) beneath what is now the Chinatown area of the city centre. Apart from several hundred metres of tunnel and storage for the equipment, the GUTE was also to act as a long term air-raid shelter and survival facility for thirty-five maintenance staff. There were also several much longer and smaller tunnels that ran out to exchanges at Ardwick (nearly 800m/2,625ft away), Dial House (900m/2,955ft away) and Salford (1.5km/1 mile away), each with a shaft to the surface.

The GUTE became operational as a working trunk exchange in 1958 and was linked to a new surface exchange (Rutherford) in 1967. A 110-m (360-ft) tunnel was added in 1972 to link the exchange directly with Irwell House exchange in Salford. Despite the cost of this infrastructure, the whole thing was decommissioned in 1988.

News of the strangest, and what would be the longest, set of tunnels under Manchester came to light following an article in a local newspaper, written by Joy Hancox and published in 1973. She received a response from former engineer William Connell, who suggested that there is a network of long-forgotten, possibly ancient Roman, tunnels that fan out from the city in at least four directions, almost at the cardinal points.

If Connell's suggestions are correct, there would be almost 40km (25 miles) of tunnel running out to Old Trafford, Crumpsal, Reddish, Wardley, Kersal, Clayton, Bradford Colliery and All Saints. Such mysteries remain to be discovered.

London

On Roman shoulders

The English capital, straddling the Thames River in the southeast of the country, is home to nine million people. It also possesses one of the world's most diverse varieties of intricate, hidden and well-used passages, ducts and tubes beneath its streets.

The reasons for this are complex. Firstly London is one of the world's longest continually inhabited regions. Secondly, it was effectively the world's most populous urban area during the first period of industrialization. Thirdly, in Victorian times, the city became the centre of one of the largest empires ever known. And fourthly, during the twentieth century, London fell under threat of invasion during both world wars. All of these factors contributed to it becoming home to many pioneering and experimental tunnelling methods, not least of which were employed in the establishment of the world's first underground railways. And despite the size of its human population being overtaken many times (by Tokyo, New York City and Beijing) it retains a global status, expanding its subterranean infrastructure to match a blossoming skyline.

Roman roots

While Paris is peppered with quarries and Canadian cities are underpinned by cavernous shopping malls, London takes the prize for sheer diversity. And compared with what the ancient Romans built in Rome – a city in which antiquity is mostly still visible above ground – some of the oldest structures beneath London date back to the city's ancient Roman roots.

Established around 43 CE, the initial Roman settlement of Londinium was sited on the north bank of the Thames River in a 1.3-km (0.8-mile) area that equates approximately to what is now known of as the City of London. Once home to at least thirty thousand people, hundreds of buildings, streets, a fort, a bath house and even an amphitheatre, the ancient city is now lost, or buried, under modern London, with a small number of walls poking up above street level. Archaeologists have so far discovered thousands of structures and artefacts, and several access points exist where the public can view remnants of Londinium below ground. Given the extent of the original Roman city and ongoing excavations, it is likely that more intriguing finds will be uncovered in the coming years.

When the Romans left Britain in around 400 CE, it's generally understood that the majority of their roads and buildings lay abandoned. The next invaders, the Saxons (400–500 CE), built an entirely new town beside the Roman remains, in the area that is now the City of Westminster. During the ensuing Anglo-Saxon and medieval period, settlements were established across the entire area now called Greater London. The vast majority of these have also been lost, with the exception of a few isolated stones and the small number of buried remains that have been discovered and are now preserved.

It was not until the ninth century that people started to reoccupy the former Londinium area, which by that time had fallen into disrepair. Stones

ABOVE The remains of a floor which was once a Roman bath house at Billingsgate. Steam would circulate between the pillars which are seen beneath. This entire site laid under the earth, unseen for centuries, until it was rediscovered in 1848 during construction of the London Coal Exchange. It was designated a protected heritage site, forming part of the first Ancient Monuments Act of 1882.

were repurposed and the city began to grow during the Tudor period, primarily as a result of river-based trading. By 1600 some 200,000 people were crammed into the area bounded by the Roman walls. Although the Great Fire of London in 1666 destroyed one-third of the city, the rebuilding tended to retain London's Roman and medieval street pattern, much to the chagrin of architects such as Christopher Wren, who envisaged a design emulating the grand baroque style adopted in France and Italy.

London's super-sewer

As the Georgian era dawned in the early eighteenth century, London's population increased from around 600,000 to about two million, making it the world's largest city (an accolade held by Beijing up until that point). It was during the Georgian expansion of London that its first forays into building structures beneath the streets really began. One of the finest surviving examples is the Marylebone Ice House, a 9m- (29.5-ft) deep well below Regent's Crescent, excavated around 1780 to store ice imported from Norwegian fjords and used for the cooling and preservation of foods for the growing elites. It's one of London's oldest deliberately constructed underground spaces.

Over the centuries, a number of small tributaries running into the Thames were either dammed to form reservoirs or channelled into conduits, many of them covered over. This process began as early as 1245 when the Bayswater Conduit piped a portion of the water from the Westbourne River between Paddington and Cheapside into the Great Conduit, which was enlarged in 1479. By then the Great Conduit also contained the Tyburn River: a commemorative plaque to it survives on Marylebone Lane, dated 1776. The Fleet River was covered over from the mid-1700s. The culverting of London's rivers did not, however overcome the stinkier problem: sewage.

A number of rudimentary brick sewers had been built in the 1600s, including those along the Wallbrook and Fleet river courses, but with the rapacious growth of water-hungry industries and around two million souls living in the capital, all with their effluent flowing into the Thames, the issue of foul-smelling, highly toxic waste in London's main river and water supply was beginning to pose a serious risk to human health.

There were several outbreaks of cholera in the 1830s–1850s, with almost 11,000 people dying of the disease between 1853 and 1854.

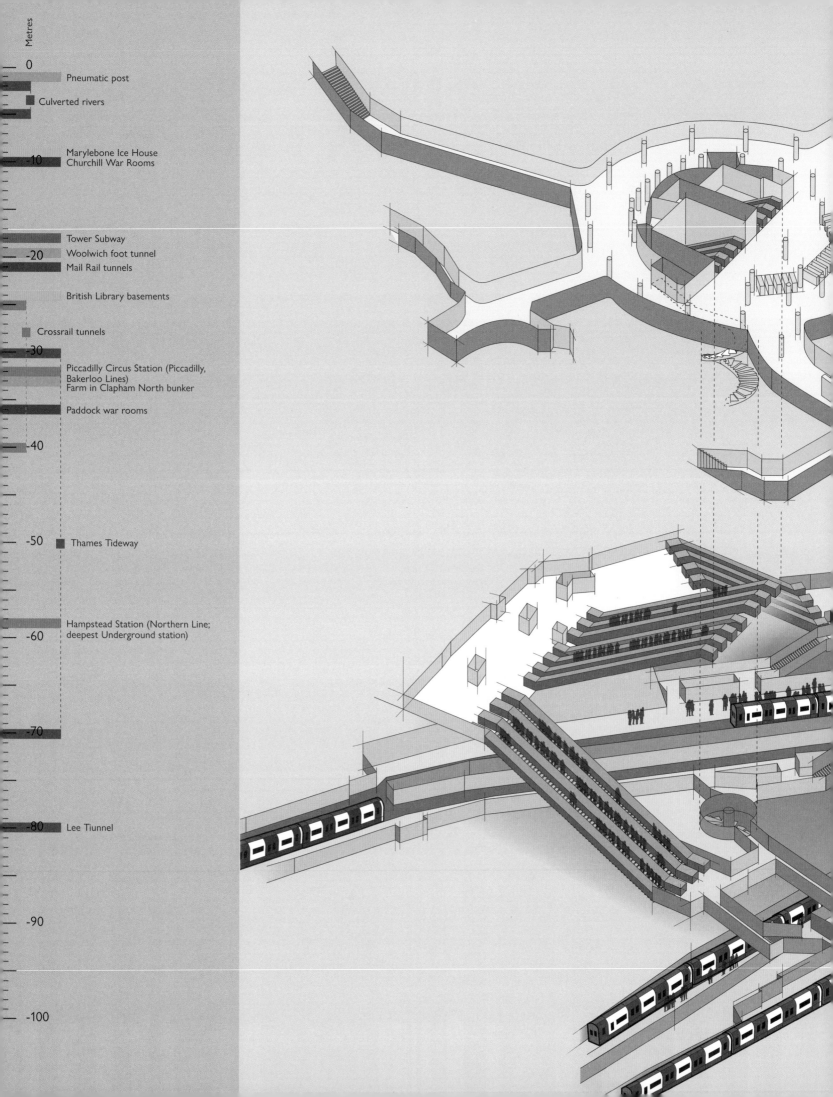

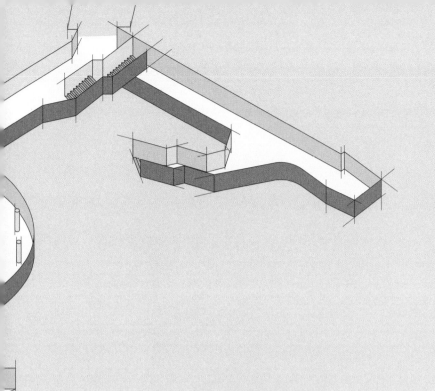

Piccadilly Circus Station

Most pedestrians passing through London's iconic Piccadilly Circus probably spend more time being mesmerized by the bright LED advertising screens than thinking about what lies beneath their feet, but under Eros is a tangled web of tunnels, both in constant use, and abandoned.

When Piccadilly Circus tube station first opened in 1906, the only way between the street and the platforms 30m (98ft) below was by lift. By the early 1920s up to 18 million passengers were squeezing into the elevators, so London Transport decided to build a new sub-surface ticket hall with eleven escalators leading to the Bakerloo and Piccadilly lines. Architect Charles Holden began work in 1925 and builders John Mowelm completed the job in three years, while the Eros statue was moved to a temporary home on the Embankment. The resulting circulating area in biscuit-coloured travertine marble and art deco pillars and uplighters was described as a 'masterpiece of opulence and chic'. The lifts and their access tunnels to the platforms were closed in 1929 but remain in ghostly empty silence beneath one of the capital's busiest subterranean spaces.

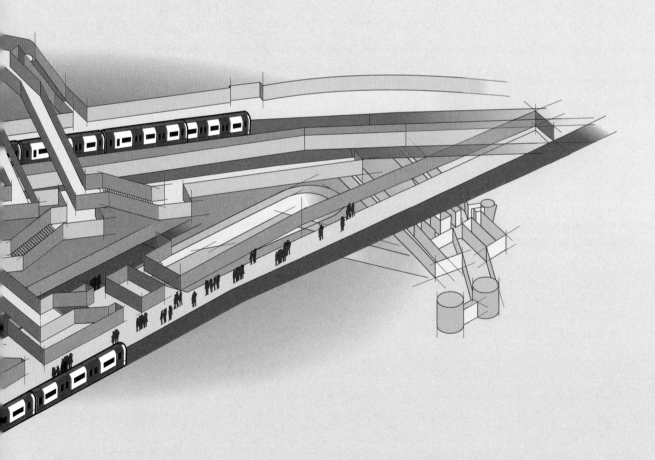

The Metropolitan Commission of Sewers appointed consultant engineer Joseph Bazalgette to develop a plan to tackle the problem. His idea for well-built local drains leading to larger main sewers and outfalls long distant from the centre of London was at first deemed too expensive. But Parliament was situated on the banks of the polluted river and when the hot summer of 1858 produced putrid aromas in an episode described as the 'Great Stink', Parliament itself was disrupted and Bazalgette's plan was rapidly approved. Over 1,800km (1,120 miles) of smaller drains would flow into 130km (81 miles) of larger, brick-lined main sewers.

The construction of Bazalgette's labyrinth took a decade to complete. By the time it reached the north bank of the Thames, plans were advanced for construction of a new underground railway. Ingeniously, the two projects were combined to form Victoria Embankment, containing both the sewer and the Metropolitan District Railway, which opened in 1870.

In recent years, owing to the age of the original Victorian sewers and the doubling of the maximum population they were intended for (four million), environmental agencies have proposed a new project to rid London of excess rainwater and to deal with the extra sewage the city is generating. Boring began on the Thames Tideway Tunnel in 2016. The new 25-km (16-mile), deep-level route will cost almost £5 billion and runs mainly beneath the river from Acton in the west to Abbey Mills in the east. It should be complete by 2023. In a nostalgic nod to the original engineer, the construction company is called Bazalgette Tunnel Ltd.

The London Underground

The real success story beneath London is, of course, the Underground. Aside from drainage, the opening of the Metropolitan line between Paddington and Farringdon in 1863 gave rise, arguably, to the most common reason to puncture the earth below our cities: mass rapid transit.

Although the northern cities of Liverpool and Manchester had a head start on building railways, England's capital wasn't far behind. By the 1840s, London was encircled by railway termini, but the

ABOVE The Metropolitan Railway company opened the world's first underground line in January 1863 and thanks to thoughtful restoration, this contemporary drawing of a steam train arriving at one of the Baker Street platforms is still recognizable today.

LONDON

biggest obstacle to crossing the great city was its historic centre: Parliament itself legislated against simply smashing up the existing urban area in order to push new railways further into central London.

City solicitor Charles Pearson suggested connecting the disparate terminals by placing a new line *below* the streets in a 6-km (3.7-mile) tunnel. Construction – complete by the end of 1862 – was constrained by the locomotive power of the day, and trains were pulled by steam-powered engines. Despite ingenious attempts to recycle their exhausts, they filled the tunnels with smoke and fumes, so they had to be built in relatively shallow trenches with plenty of gaps for letting out gases. Despite dire warnings of asphyxiation in the sooty atmosphere, the initial line was so successful it was soon extended and spurred the development of additional tunnelled railway lines.

The Metropolitan District Railway filled the space above Bazalgette's Victoria Embankment sewer, leading to the completion of a full circular route by 1884. Within a few short years, following the discovery of electricity, it became possible to power trains using this newfangled source of energy. The first London tunnel to benefit from it was the City and South London Railway, which opened in 1890.

BELOW London's newest subterranean stations on the Elizabeth Line have been designed with simplicity, safety and longevity in mind: smooth sweeping curves in this cross passageway at Farringdon station were designed by building consultancy practice AHR.

ABOVE Situated beneath the building that now houses HM Treasury (on the corner of Great George Street and Horse Guards Road) are some of Britain's most crucial underground rooms. With war looming in 1938, a large space was excavated to house Government meetings and provide a central command operation centre for the Prime Minister. The facility hosted over 100 Cabinet meetings and Churchill directed operations from the Map Room and via transatlantic calls to the US president. Converted to a museum in 1984, and now known as the Churchill War Rooms, it is run by the Imperial War Museums.

It also became feasible to bore deeper tunnels than the subsurface lines that had skirted built-up London to this date. Built using this method, the Central London Railway was opened in 1900. Known as 'the Tube', a name that is now commonly used for the whole network, it ran between the capital's busiest shopping areas around Oxford Street and the City, near St. Paul's Cathedral. The bright, white-tiled platforms and cheap 'tuppenny' fare increased interest in providing more lines.

Three additional routes, the deep-level tubes that would penetrate and properly serve the West End, opened between 1906 and 1908 and eventually became known as the Bakerloo, Northern and Piccadilly Lines. The combined tunnelled rail services were collectively named the Underground from 1907. Such advances in engineering spurned similar ventures in Budapest (1896), Glasgow (1896), Paris (1900), Berlin (1902) and New York (1904).

Following a spurt of building in the 1930s and another just after the Second World War, construction on the Victoria Line began in the 1960s and the Jubilee Line in the 1990s. Underground construction was back on the agenda in the first quarter of the twenty-first century, with the Crossrail project that began in 2009. Boring 21km (13 miles) of new tunnels beneath London, the Elizabeth Line, as it as named, carries mainline-sized trains from both the eastern and western suburbs and is comparable to the Paris RER system.

Unique feats of engineering

Although the Underground dominated tunnelling below London, there are several other subterranean developments worthy of note. For example, there was the incredibly ambitious Thames Tunnel between Rotherhithe and

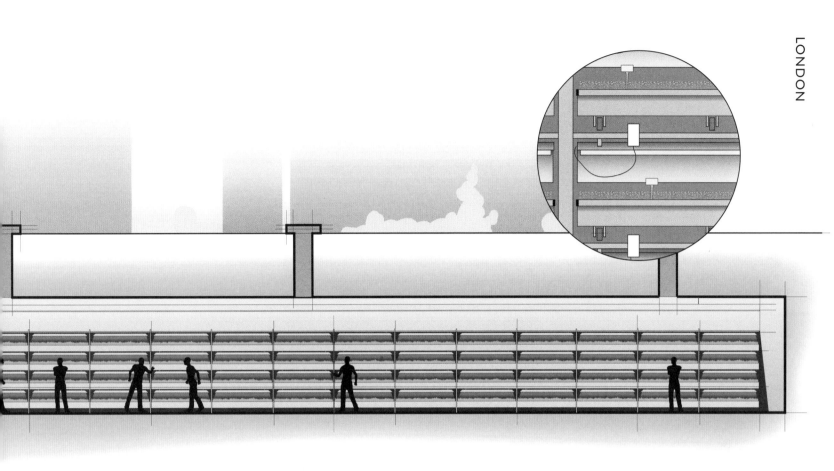

ABOVE Following the 1940 Blitz of London, plans were rapidly advanced for ten large-capacity air raid shelters, each of which was supposed to protect 10,000 people. Though two were not pursued owing to poor ground conditions, eight were constructed in just two years. One such shelter was built under the Northern Line station Clapham North. In 2014 an enterprising Zero Carbon Food company bought the space to be converted into an underground urban farm. Their hydroponic system now produces fresh micro greens and salad leaves, 33m (108ft) below Clapham's streets.

Wapping. Running up to 23m (76ft) below the river at high tide, this bold project took nearly two decades and two generations of great engineers to complete. Originally proposed for use by horse-drawn carriages, work started in 1825 using Marc Isambard Kingdom Brunel's tunnelling shield. Following several accidents and deaths, not to mention rising costs, the under-river crossing was finally completed by 1843, but despite its grandeur it was not as popular as hoped and was converted to rail in 1869. It is still in use today, as part of London's Overground network.

Equally beguiling was the Tower Subway, a pedestrian tunnel running from Southwark to a spot beside the Tower of London. Opened in 1869 it was later converted to a cable-hauled railway only to be deemed unprofitable and closed for that use just months later. It continued to serve pedestrians for a short while before that, too, ceased to be profitable and was closed less than thirty years later. This great example of Victorian engineering innovation now conveys little more than water mains.

Other notable subterranean structures in London include: the pneumatic messaging network, a system of delivering post by pressurized air tubes (1853); the London Silver Vaults (1885); the Woolwich foot tunnel (1912); the Post Office Railway (a.k.a. Mail Rail, 1927–2003, but now open to public visits); 'Paddock', a secret communications centre in Dollis Hill (1939); Churchill's War Rooms (1939); Oxgate Admiralty Citadel (1940); the deep-level express tube (under the current Northern Line and used as an air-raid shelter but never converted to rail use after the war); and the Kingsway trunk telephone exchange (c. 1945).

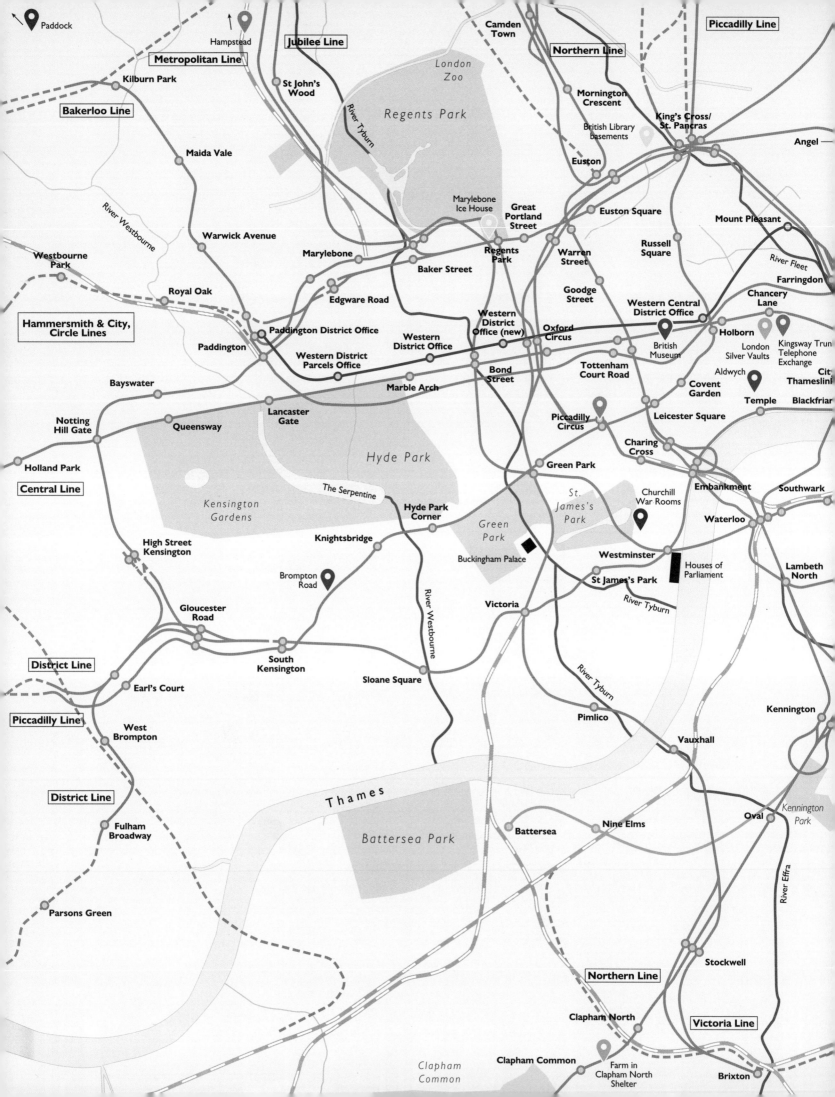

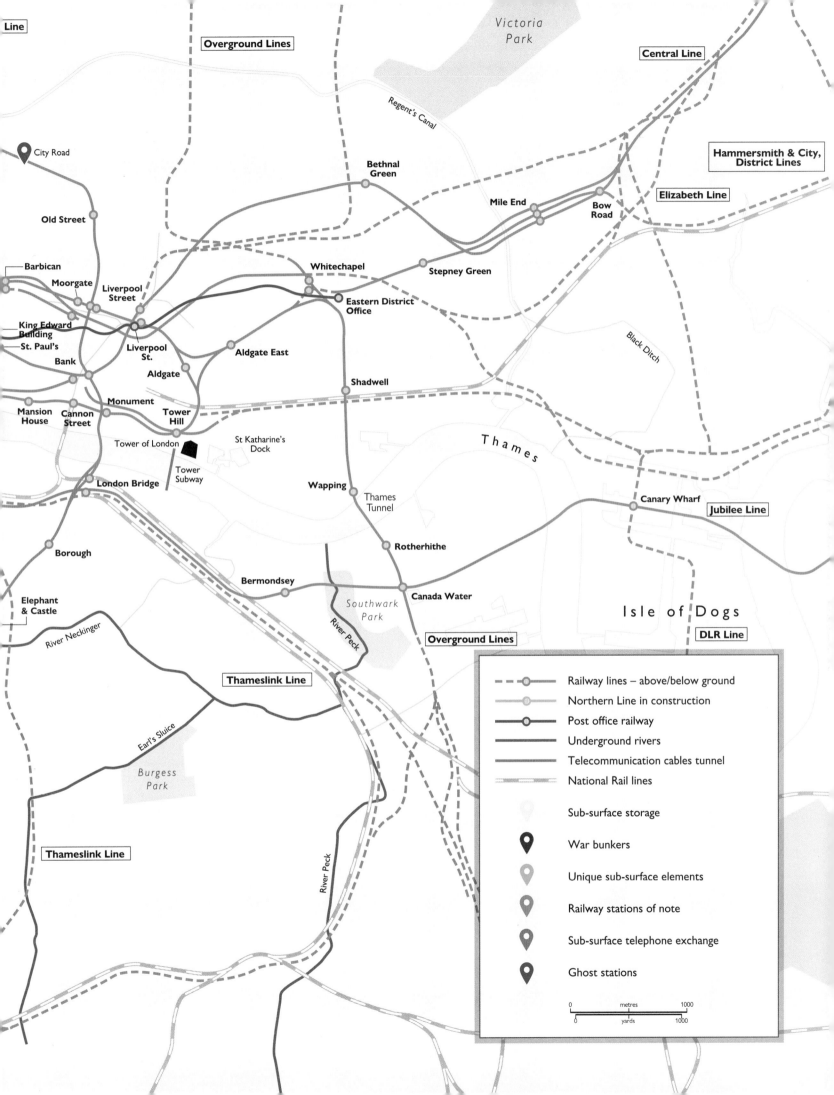

Barcelona

The planned city

With a current population of 1.6 million people and five million in the wider metropolitan area, Catalonia's largest city – and Spain's second biggest – originated beside the Mediterranean Sea in ancient Roman times. It was the Visigoth capital from 507 to 573 CE, and the centre of its own county after merging with the Kingdom of Aragon in the Middle Ages. It eventually became the capital of Catalonia Principality and is now one of the key cultural and tourist beacons of Europe.

Ancient roots

The Roman presence in Barcelona dates back to 218 BCE when Barcino, an extensive walled settlement of approximately 2,000 people, was established as a satellite of Rome itself. As with a number of ancient Roman cities, many of the original buildings were covered over and are said to remain beneath the modern streets. Some claim that Barcelona is home to the world's most extensive subterranean Roman ruins, parts of which are visible today beneath the Plaça del Rei and can be accessed via the Museu D'Història de Barcelona, the city's museum of history. The surprisingly detailed artefacts down there include everything from a textile dye works to a production facility for wine. There are also some surviving Roman sewers and water supplies in the area – for example, a long stretch of aqueduct, understood to be from around 20 CE, has been unearthed at Badalona.

When the Roman Empire broke up in the fifth century CE, urban growth continued outside the old centre of Barcelona, later called the Barri Gòtic, or Gothic Quarter. The city underwent a big expansion during the eleventh century, in an area known as El Born, and again between the thirteenth and fourteenth centuries, when El Raval was built. New defences were erected around 1285 – in what was then known as Barchinona – during battles against the French, and these were reinforced in later decades. By this time the city was full and small settlements known as *vilanovas* sprung up outside the walls. It was not until the fourteenth century that the Rambla wall was constructed, and it took until the 1850s for the city to officially escape the confines of its old walls.

Urban planning

As the Industrial Revolution swept through Europe in the late nineteenth century, there was suddenly a great demand for space, but Barcelona's walls confined the city to an area in which there was no room for new factories, homes and railways. A competition to design a vast expansion of Barcelona was won by a civil-engineer-turned-urban-planner, called Ildefons Cerdà. His concept of 'islands' of up to 900 apartment blocks on long avenues and cross streets (known as the 'grid') is now called the Eixample district. It was a stark contrast to the historic layout of narrow streets in the Roman and medieval parts of the city.

The blocks within Eixample – called *manzanas* – were originally planned to have just two or three sides overlooking an enclosed interior space of parkland or gardens. Cerdà also planned an

ABOVE The bases of these once mighty columns of the Forum and Basilica are now many metres below the Plaça del Rei. This area is part of what became the first section of the Museu d'Història de Barcelona (MUHBA), which opened in 1943. Visitors can now stroll the pavements of the Roman settlement which are located immediately beneath the bustling twenty-first century square above.

underground utility supply, which was gas in those days, and the sewers. At intersections, the blocks had chamfered corners, enabling a turning curve suitable for future tram lines, a feature they retain to this day.

As this neatly planned extension marched across the plain, swallowing up the *vilanovas*, its rigidity inspired other architects to rebel. Among them, Antoni Gaudí created lush, flowery ornamentation for the apartments he designed and embarked upon the never-ending project of the Sagrada Família church.

The Barcelona Metro

The first railway line on the Iberian Peninsular opened in 1848 between Barcelona and the nearby coastal town of Mataró. A short rapid transit line, Ferrocarril de Sarrià a Barcelona, with four underground stations, opened in 1863, but a proper metro system did not arrive until 1924, when the Gran Metropolitano de Barcelona began operating between Lesseps and Plaça de Catalunya. The Metro continued to expand, albeit slowly over the coming years.

An odd quirk of the system is that there are a number of closed and resited stations, as well as a couple that never opened at all. Among these 'ghost' stations – a dozen in total – are the following: Banc (1911), which never opened but would now be on L4; Bordeta (1926), closed in 1983 owing to its proximity to the Santa Eulàlia station; Correos, a 1934 station

Metres

0 — Aqueduct at Badalona

-10 — Gaudí Station / Diamond Square Air-raid Shelter Refugi 307

Joan Miró storage tank

-20

Desalination plant

-30

-40

-50

-60

Llefià Station (Line 10)

-70

El Coll / La Teixonera Station (Line 5; deepest Metro station)

-80

-90

-100

Llefià Station and Diamond Square Air-raid Shelter

Catalonia has a proud tradition of innovative and experimental art and design and this can be seen both above and below ground. Nowhere is this more evident than in the Barcelona Metro system which has been expanded exponentially in the last two decades.

The new L9 and L10 currently under construction will be the deepest lines in the city (up to 80m/262.5ft down in some places) and will share a common central section which is being built with branches at either end. The lines will eventually run to almost 50km (31 miles). Both L9 and L10 have sections open on either side of the city (currently operating as L9 Nord, L9 Sud, L10 Nord and L10 Sud). For technical reasons (unlike other Barcelona Metro lines), the tunnel boring machines excavated a 12-m (39.4-ft) diameter tunnel so that one train can run on top of the other.

Historical Catalonian design can also be explored in the vast number of air-raid shelters beneath Barcelona, some of them with complex tunnel networks, built with care to protect its inhabitants.

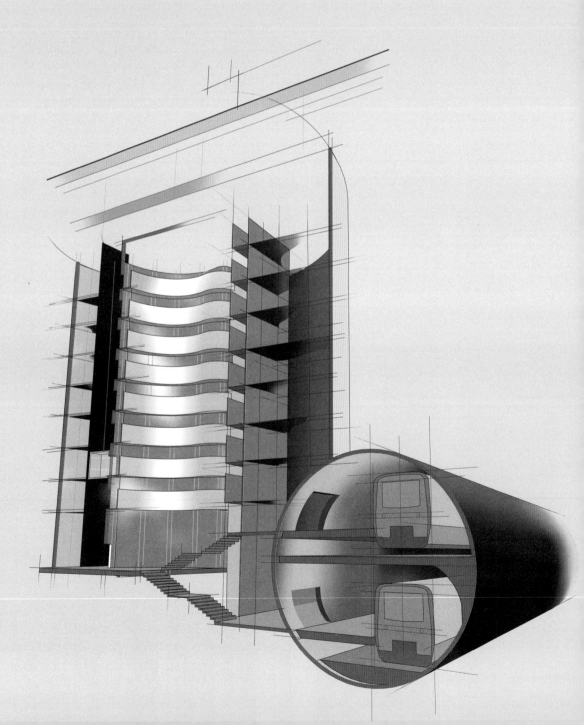

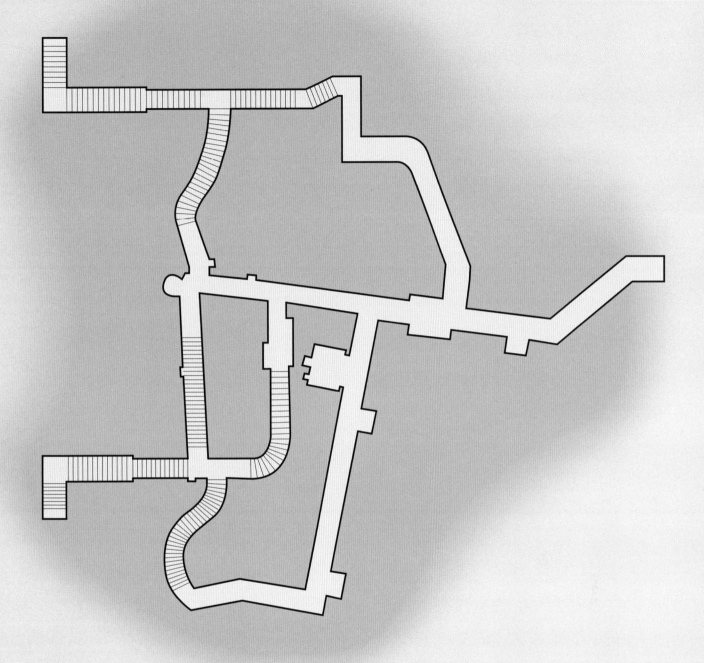

LEFT On the L10 Nord between Sagrera and Gorg, two neighbouring stations (La Salut and Llefià) have received especially notable treatment. La Salut lies at 30m (98ft) beneath the surface, and Llefià at 63m (207ft). Both are reached by a deep well 25m (82ft) in diameter. In order to reduce any feelings of vertigo or claustrophobia for passengers going between surface and platform, architect Alfons Soldevila Barbosa made extensive use of stylish horizontal features with lighting and walkways.

ABOVE Lying 12m (39.4ft) beneath Plaça del Diamant Square (Diamond Square), this is one of the largest and best-preserved air-raid shelters built during the Spanish Civil War. Capable of protecting around 200 people, its 250-m (820-ft) network of tunnels extends beneath the Square and towards both nearby Streets Carrer de les Guilleries and Carrer del Topazi. Rediscovered in 1992, it was made open to public tours in 2006.

that became a victim of an L3 extension in 1972; Fernando opened in 1946 but closed in 1968 for similar reasons as Correos; Gaudí (1968) never opened because of its proximity to the Sagrada Família stop next door; and Travessera, which would have been located between Diagonal and Fontana, but works were never completed. The three stations that have moved from their original locations are Espanya, Santa Eulàlia and Universitat, although their original platforms remain underground.

Preservation measures

During the Spanish Civil War, the Nationalists chose Barcelona as a testing ground for attacks. Caught unawares, the authorities had made no preparations, so the local population was forced to dig bomb shelters. Some 1,000 shelters were built during this time, almost entirely by hand. One recently unearthed in the Poble-sec district, Refugi 307, has more than 400m (1,300ft) of tunnelled corridors housing utilities such as bathrooms, a water fountain and even a fireplace. This tunnel network gave shelter to thousands during the many bombing raids visited upon the city.

Given Barcelona's location – on a plain between mountains, the Llobregat River and the sea – it does experience a lot of water run-off during inclement weather. The area suffers from occasional flash flooding, which prompted the authorities to create a network of eight underground and two open-air storm overflow collection tanks. Constructed in and around the city since 1997, these can handle almost 50,000m^3 (1.77 million ft^3) of water. The Catalonians are justifiably proud of them, despite their expense, as they have actively prevented further inundations.

One of the largest collection tanks is located in the Univeristat area. Another, named after the Spanish painter and sculptor Joan Miró, has a surreal, almost sci-fi feel, with huge concrete pillars. The filtration system that the water passes through before it is gently released back into the environment prevents almost 1,000 tonnes of polluted suspended matter pouring into the Mediterranean Sea each year.

ABOVE The platforms of Llefià Metro station on L10 are accessed via lifts or this illuminated walkway in the well 63m (207ft) from the surface.

TOP RIGHT Exposed concrete is clearly visible at the entrance, demonstrating the purpose of this civil defence anti-air raid shelter.

BELOW RIGHT Running from the opening in 1924 until withdrawn from service in 1991, these beautiful carriages were restored to their original condition and are now run on the system on special occasions to celebrate the history of the Metro.

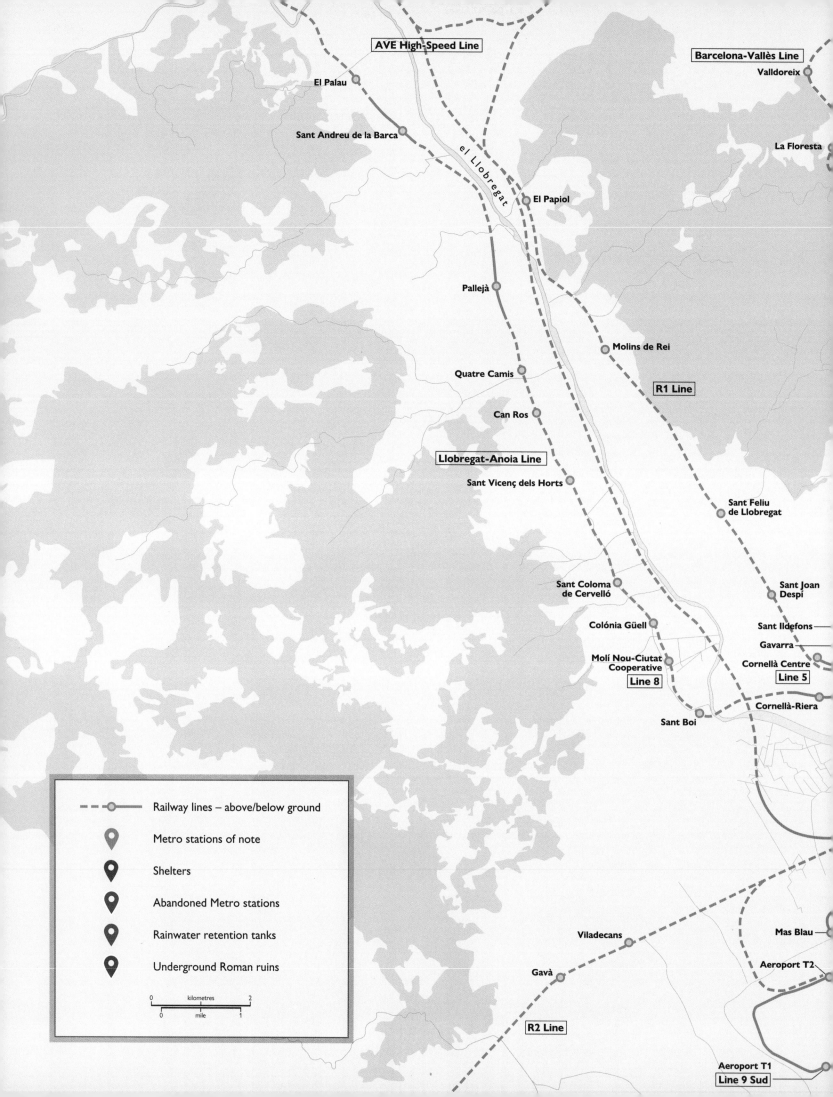

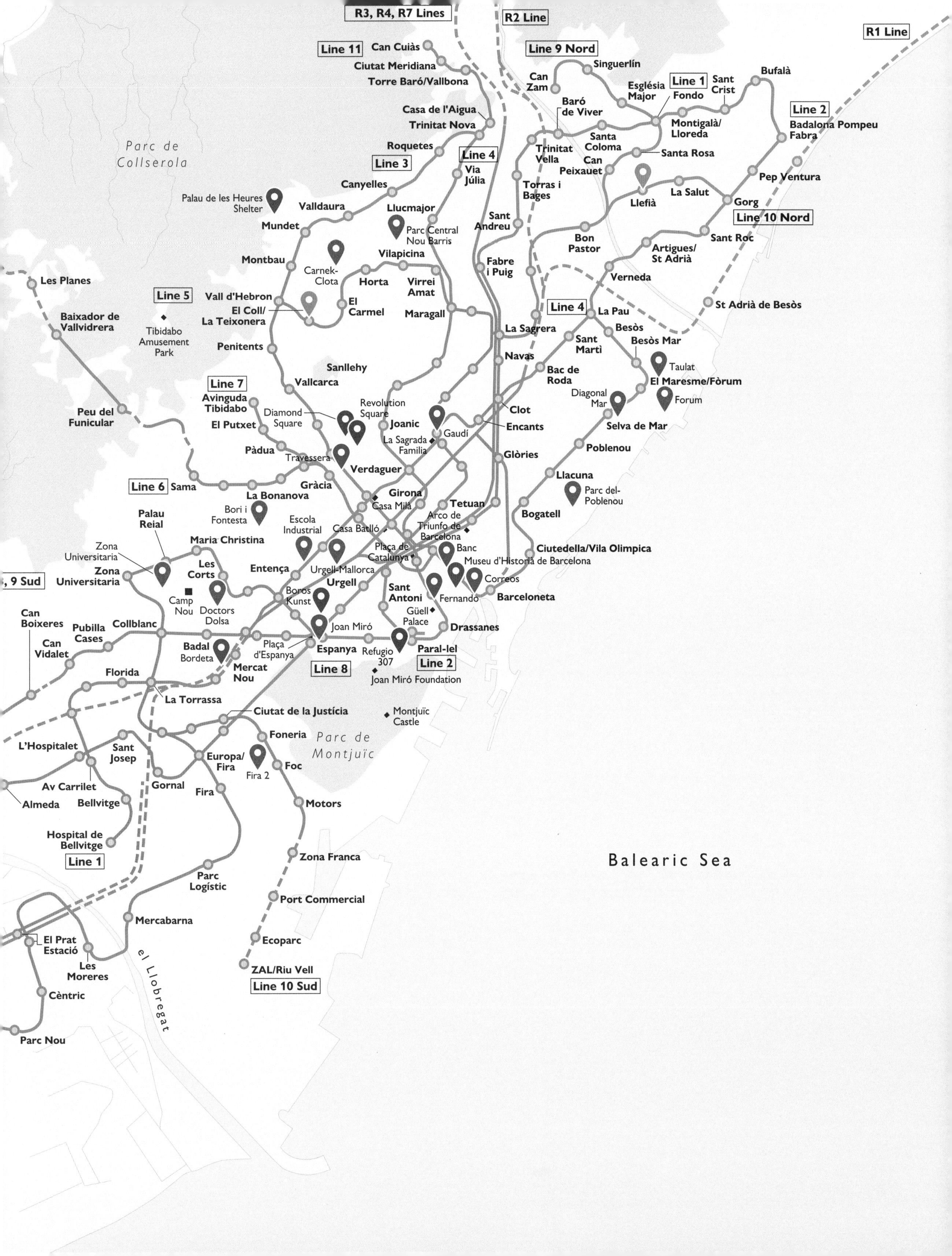
R3, R4, R7 Lines
R2 Line
R1 Line
Line 11
Can Cuiàs
Ciutat Meridiana
Torre Baró/Vallbona
Casa de l'Aigua
Trinitat Nova
Roquetes
Line 9 Nord
Singuerlín
Can Zam
Baró de Viver
Església Major
Line 1
Fondo
Sant Crist
Bufalà
Line 2
Badalona Pompeu Fabra
Montigalà/ Lloreda
Santa Coloma
Trinitat Vella
Can Peixauet
Santa Rosa
Parc de Collserola
Line 3
Line 4
Via Júlia
Torras i Bages
Llefià
La Salut
Pep Ventura
Canyelles
Palau de les Heures Shelter
Valldaura
Llucmajor
Sant Andreu
Gorg
Line 10 Nord
Mundet
Parc Central Nou Barris
Bon Pastor
Sant Roc
Artigues/ St Adrià
Montbau
Carnek-Clota
Vilapicina
Fabre i Puig
Verneda
Les Planes
Horta
Virrei Amat
Line 5
Vall d'Hebron
El Carmel
St Adrià de Besòs
El Coll/ La Teixonera
Maragall
La Pau
Baixador de Vallvidrera
Tibidabo Amusement Park
La Sagrera
Sant Martí
Besòs
Besòs Mar
Penitents
Navas
Taulat
Sanllehy
Bac de Roda
El Maresme/Fòrum
Line 7
Vallcarca
Diagonal Mar
Forum
Avinguda Tibidabo
Revolution Square
Clot
Diamond Square
Peu del Funicular
El Putxet
Joanic
Gaudí
Encants
Selva de Mar
La Sagrada Familia
Pàdua
Travessera
Glòries
Poblenou
Verdaguer
Llacuna
Line 6
Sama
Gràcia
Parc del-Poblenou
La Bonanova
Girona
Casa Milà
Tetuan
Bogatell
Palau Reial
Bori i Fontesta
Escola Industrial
Arco de Triunfo de Barcelona
Casa Batlló
Maria Christina
Zona Universitaria
Plaça de Catalunya
Banc
Ciutedella/Vila Olimpica
Museu d'Història de Barcelona
Les Corts
Entença
Urgell-Mallorca
Correos
9 Sud
Zona Universitaria
Urgell
Sant Antoni
Fernando
Barceloneta
Camp Nou
Doctors Dolsa
Boros Kunst
Can Boixeres
Pubilla Cases
Collblanc
Güell Palace
Joan Miró
Drassanes
Can Vidalet
Badal
Bordeta
Plaça d'Espanya
Espanya
Refugio 307
Paral-lel
Mercat Nou
Line 2
Florida
Line 8
Joan Miró Foundation
La Torrassa
Ciutat de la Justícia
Montjuïc Castle
Foneria
Parc de Montjuïc
L'Hospitalet
Sant Josep
Europa/ Fira
Foc
Fira 2
Av Carrilet
Gornal
Fira
Almeda
Bellvitge
Motors
Hospital de Bellvitge
Line 1
Zona Franca
Balearic Sea
Parc Logístic
Port Commercial
Mercabarna
El Prat Estació
Ecoparc
Les Moreres
el Llobregat
ZAL/Riu Vell
Line 10 Sud
Cèntric
Parc Nou

Paris

The Swiss cheese of Europe

From the Eiffel Tower to Notre Dame, the Louvre to the Arc de Triomphe, France's capital barely needs an introduction. With 2.1 million living in the city and up to seven million in the immediate metropolitan area, this historic site also lies at centre of the Île-de-France, one of Europe's most populous and prosperous regions.

A Celtic tribe called the Parisii are known to have inhabited the basin around a clump of small islands in the Seine River around 3000 BCE. The ancient Romans established a settlement on the southern (or 'left') bank of the river across from what is now called the Île de la Cité, in 52 BCE. Their town was called Lutetia Parisiorum and was substantial enough to contain an amphitheatre, along with the traditional Roman temples, baths and markets. By the time the empire had waned and Christianity had arrived, the Latin name had been shortened to Parisus. The Franks arrived from Gaul and following various fires, fortifications and attacks – by the Vikings, among others – Paris, now centred on the 'right bank' and Île de la Cité, had established itself as the largest settlement in France and a hub of art, culture, politics, religion and education by the eleventh century. The Notre Dame Cathedral, the Louvre Palace and the embryonic university were all established during this medieval period. By 1328 Paris was the biggest city in Europe with 200,000 crowded inside its walls, a figure that doubled by 1640 and reached well over half a million by the mid-eighteenth century. After the revolution of 1789 the population dropped but was growing again under Napoleon in the 1800s. By this time industrialization was advancing and the first railway was built from Saint-Lazare to Le Pecq (a royal estate on the western fringes) in 1837.

Parisian quarries

Instead of bringing stone from far away, the Parisians excavated the building blocks of their fine city from beneath their feet: thousands of tonnes of sedimentary Lutetian limestone and gypsum were hewn from the *carrières de Paris* (quarries), many of them south of the river. The excavations, begun in the 1400s, have been expanded over centuries, and are known to extend for up to 300km (185 miles), making Paris the most heavily mined urban area on the planet. While some of the vast network is accessible on official walking tours, much of it is dangerous and unlit; the 'cataphiles' (urban explorers of the city's mines) claim there are more to discover.

The quarries are also home to one of the more macabre tourist destinations in the city: its catacombs. In the late 1700s Paris suffered a number of public-health crises. Disease was rampant and the cemeteries full to overflowing. The administration decided on a radical plan to dispose of its dead: it would hide them in the hundreds of kilometres of quarries. At the time, the Tombe-Issoire at Montrouge was just outside the city limits and excavations were made here to house the bony contents of the largest cemetery, Saints-Innocents. The site was consecrated as the municipal ossuary in 1786 and opened to public viewing by appointment from 1809. Given that almost the entire contents of four major cemeteries were dug up and transferred to the ossuary, it is not surprising that there may be up to six million individuals whose remnants have been stacked and neatly arranged in the catacombs. In the last century, the site has also witnessed alternative uses as a Nazi bunker and a clandestine cinema hosting film-noir screenings.

RIGHT Built between 1868 and 1873, the Réservoir de Montsouris in the 14th arrondissement was designed to supply most of the southern half of the city with clean water. Its elegant 1,800 support pillars, blue ceramic floor and ornate iron and steel glazed turrets above ground give an air of some extraordinary covered piscine. The water is sourced from fresh springs and transported along aqueducts that run in to the city from up to 130km (80.8 miles) away and the Reservoir is capable of holding over 200,000 m³.

Floods and clean water

Following a millennium of continual occupation, Paris has an underbelly pockmarked with a myriad of trenches, drains, passages and pipes. One of the biggest problems urban planners faced was that of water. The 540-km (335-mile) Marne River flows into the Seine just southeast of the city and, with the exception of a few hills – the highest being Montmartre and Telegraphe both almost 130m (430ft) – Paris is relatively flat. Effectively it is built on a floodplain and when the Seine is in full flow it bursts its banks.

Early attempts at flood prevention led to two of the city's four islands being joined to the embankment and an ancient curvaceous arm of the river blocked up and dried out. This section was later replaced, in part, by the Canal St. Martin. Opened in 1825 the 4.6-km (2.9-mile) waterway shortened the journey along the Seine by 25km (15.5 miles). It also connected to the Canal de l'Ourcq and played a role in supplying 'fresh' water to the city. When the section of Canal St. Martin between Temple and Bastille was covered over in the mid-19th century, it created the longest tunnel in Paris. Although the river embankments were raised, the Seine refused to be tamed, culminating in the Great Flood of 1910, which penetrated the Métro system and backed-up the sewers, identifying a need to create better flood defences east of the city.

Although Paris had some of the earliest paved streets (laid around 1200) and rudimentary attempts at sewers – some ancient Roman and an open-air one built in the rue Montmartre in 1370 – the city had to wait until the seventeenth century before gaining its first underground wastewater ring and the canalization of a small Seine tributary called the Bièvre River. For a substantial vaulted, subterranean sewer system to serve the whole city, Paris waited once more until the reign of Napoleon Bonaparte in the early 1800s, when 30km (18.5 miles) of them were constructed.

It was Napoleon III who employed city planner Baron Haussmann, who in turn engaged the engineer Eugène Belgrand to construct a 600-km (370-mile) system of brick-lined sewers and drains, much of which was designed to lie beneath the 200km (125 miles) of new pavements and boulevards Haussmann masterminded.

Between the end of the First World War and the mid-1970s, another 1000km (620 miles) of sewers were connected to the existing system beneath Paris, a techinical achievement that is celebrated in the Musée des Égouts de Paris (Sewer Museum) in the city's 7th arrondissement.

The Paris Métro

The Parisians were acutely aware of the industrial advances being made across Europe, particularly in Germany and the United Kingdom. As in London and elsewhere, the arrival of railways in Paris in the 1830s happened well after the city was established and, therefore, their stations were inevitably sited

at the outskirts of the city, forcing passengers from the provinces to disembark and traverse the busy streets with their luggage in horse-drawn trams or cabs, or on foot for a connection to another train.

The mainline railway companies had long sought a method of connecting their distant terminals, and the Parisians were getting tired of long walks, overloaded trams and expensive cabs. Several ideas, many of them ahead of their time, were considered for rails in tunnels, including an 1854 concept by Édouard Brame and Eugène Flachat for a 2.2-km (1.4-mile) freight railway below the street from Gare du Nord to Les Halles markets. Other ideas of the era were far more fanciful and all were rejected.

During the early 1860s, newspaper announcements that London was building the world's first urban railway in a tunnel had a galvanizing effect, both inspiring and infuriating the French. Rivalry is a great driver of creativity. Within months, a wealth of proposals surfaced hoping to equip Paris with some form of rail-based mass transportation across their beloved city. Elevated railways along the lines of those in New York were ruled out, and a host of well-respected engineers advocated various types of underground transport. But a rather French political standoff ensued between the national interests of the State supporting the mainline railway companies and the city administrators, who favoured a system designed for more localized connections. A plan by engineer Jean-Baptiste Berlier in 1887 proposed three *tubulaire* routes, with trains running in iron-clad tubes just below the surface.

TOP LEFT The construction of the first line of the Métropolitain was achieved almost entirely by the 'cut'n'cover' technique where the road was dug up, and a trench created by shoring up the sides. The street surface was then reconstructed over what in effect became a tunnel. It caused chaos on major thoroughfares like the Rue de Rivoli, seen here in May 1899.

BOTTOM LEFT To breach the River Seine and waterlogged ground around it for Line 4, an entirely different method was required: the 'tunnels' were built from steel on dry land like here at Saint-Michel, then either floated into the river and sunk to rest in trenches dredged from the surface or using compressed air (at great danger to the workers) the ground beneath was dug out and the metal structure lowered into the cavity created below.

Deadlines set by big events in the city came and went, including the Exposition Universelle of 1889, for which Gustave Eiffel constructed his eponymous tower, but still there was no progress on a metro system. Another wake-up call came in 1895, with the short extension of a suburban line (the Ligne de Sceaux) in a tunnel that nudged two stations closer to the centre. A year later, with the Exposition Universelle of 1900 looming, the city council finally signed off civil engineer Fulgence Bienvenüe's plan to build a mainly subterranean ten-line network

In the end it was a classic compromise: the system would link some of the mainline stations (gares de l'Est, du Nord, du Lyon, du Montparnasse and Saint-Lazare), plus provide lines for local people to cross their city. Construction began in 1898.

The first line to be completed, all in tunnel, was the key southeast–northwest axis along the right bank of the Seine. It had to be built using the cut-and-cover method, which caused mayhem on the busy Champs-Élysées, Rue de Rivoli and Rue St. Antoine. There were also two small branches that would later form the basis of other lines. The chemin de fer Métropolitan opened in 1900, just in time for the exposition, but the only mainline station served was the Gare du Lyon. The platforms and passageways were lined with brilliant white ceramic tiles – a feature that remains to this day.

A second line curved in a semicircle around the northern boulevards (the trajectory of the old city walls) and included four elevated, open-air stations. Although it ran close to several mainline rail stations it was hardly handy for any of them. A third line did at least serve Gare Saint-Lazare and a line around the southern boulevards – also with elevated sections and serving twelve open-air stations – gave access to the stations at Montparnasse and Austerlitz.

The crucial gares du Nord and de l'Est took a bit longer to reach. The Métro, as it quickly became known, was an immediate success with Parisians, and the initial system was largely completed within just a decade. A complimentary network of three additional routes was also agreed, with the Nord–Sud lines opening in 1912. The larger sections of the Nord–Sud lines used bored tunnels and had decor on the entrances. A circular ticket hall beneath Saint-Lazare still survives and is a symphony of beauty in faience.

Métro curiosities

Several anomalies have created fascinating underground spaces in the Paris Métro. Firstly, most lines have no branches. Large turning circles constructed at their terminals enabled trains to face the opposite direction for the return journey. Though some of these are still used, most have been bypassed by extensions further out and are used as parking garages. Secondly compared to other cities, the stations were built much closer together – it's often possible to peer down the tunnel and see the next brightly lit platform in the darkness. Thirdly navigating past kilometres of old quarries caused many engineering challenges. Line 7, for example, passes through several large cavities beneath the heavily mined Buttes-Chaumont. At one point, where the tunnel opened into a huge void, pillars were built from the cave floor to support an enclosed tube of stone and steel through which trains could pass.

Elsewhere in the system, the terrain was so difficult that the two running tunnels had to be stacked on top of each other as opposed to side by side. In addition, whole sections were part constructed but never opened. For example, in the 16th arrondissement several tunnels link Lines 9 and 10 but have never seen a train in public service. There is even a station complete with island platform but no access to the street above. Métro history fans of the Association d'Exploitation du Matériel Sprague (ADEMAS) sometimes hold tours and even Christmas dinners on the platform.

Paris has many ghost stations that were either opened then closed or built at platform level but never finished off. Haxo is a good example of this. Situated on a rarely used connecting tunnel called la voie des Fêtes that runs between the short shuttle lines 3bis and 7bis the platform at Haxo was built as the only intermediary stop between Place des Fêtes and Porte des Lilas. It was never connected to the surface, however, and has since been used only to test new designs. Other long-closed stations include Arsenal, Champs de Mars, Croix Rouge and Saint-Martin, all shut during the Second World War owing to their proximity to others. The last, Saint-Martin is worth seeing on rare ADEMAS visits, as its full of sculpted plaster advertisements, once a common sight, now all obliterated elsewhere.

A couple of closed stations are visible. Get off at Gambetta on Line 3 and walk back along the platform (towards Paris). The platform is unusually long because it was once part of Martin Nadaud station, which closed when it was merged into Gambetta, 230m (755ft) along the track, in 1969.

The concrete boxes for two proposed stations were built near La Défense in the 1930s. Plans for that line changed and they were never used, but sit empty beneath new surface buildings. There is a similar unused concrete box beneath Orly-Sud. Other peculiar stations

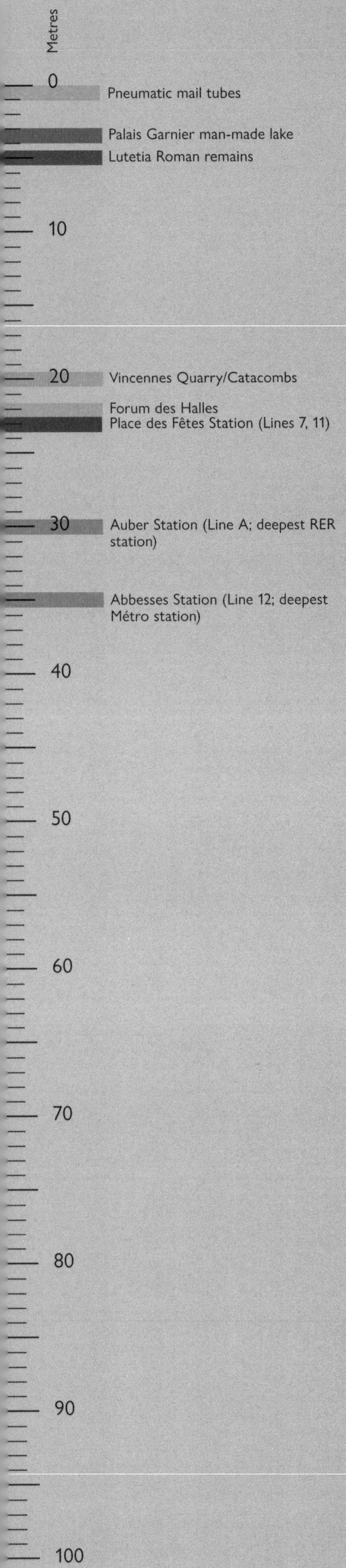

Metres

0 — Pneumatic mail tubes
   Palais Garnier man-made lake
   Lutetia Roman remains

10

20 — Vincennes Quarry/Catacombs
   Forum des Halles
   Place des Fêtes Station (Lines 7, 11)

30 — Auber Station (Line A; deepest RER station)

   Abbesses Station (Line 12; deepest Métro station)

40

50

60

70

80

90

100

Forum des Halles and Municipal Ossuary

Paris is the home of the world's largest ossuary. A long established Roman and Orthodox Church tradition, the word means to exhume the ancient corpse, salvage the bones themselves and place them in a small box or, in this case, arrange them in the catacombs.

A beacon of twenty-first century Paris shopping is Forum des Halles: the site was occupied by the city's main fresh food market, Les Halles (which began in the twelfth century), but was relocated to the suburbs and demolished in 1971. At the same time the deep-level express metro to be called RER was seeking a central location for its new interchange station. An enormous cavern was dug out below the old market (called *le grand trou*) and space was provided for the crossing of two new RER lines with access to four (later five) Métro lines. Atop the trains a modern shopping mall, cinemas and swimming pool were constructed, largely underground. A major reconstruction of the mall was undertaken in 2010 and inaugurated in 2018.

BELOW The transit hub Châtelet–Les Halles now sees 750,000 passengers passing through it every day, making it Europe's busiest underground rail station; whilst the shopping centre gets up to 150,000 daily visitors.

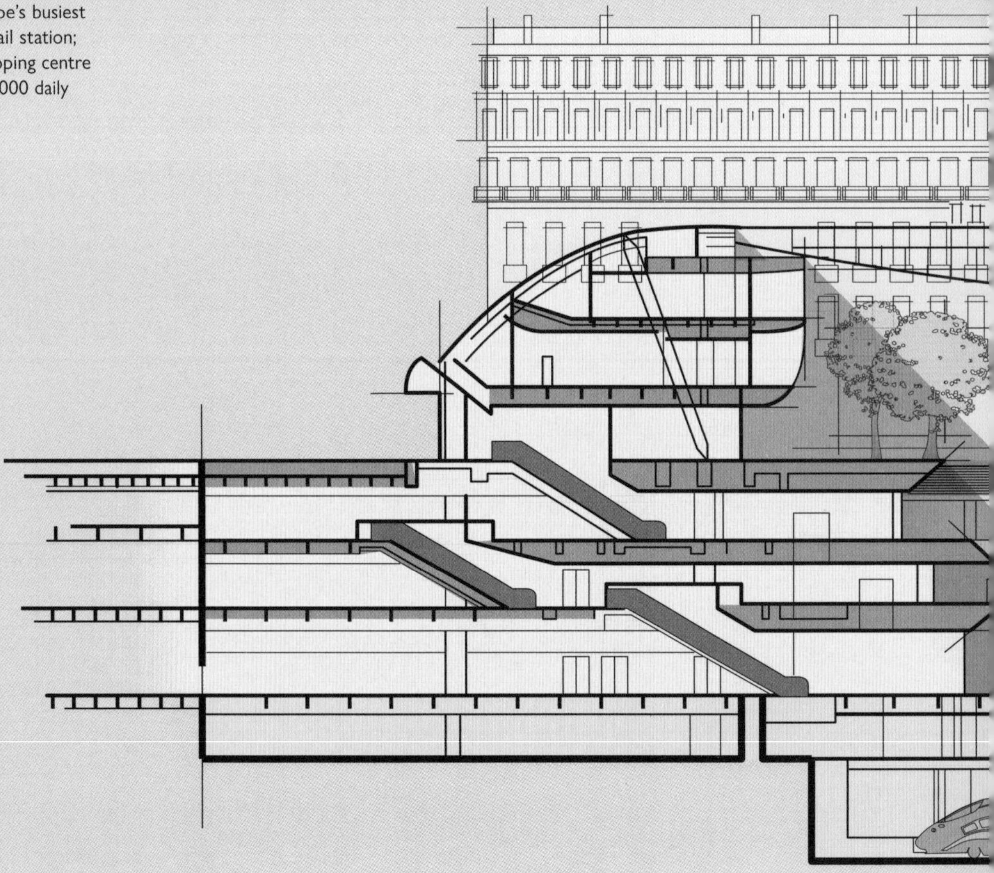

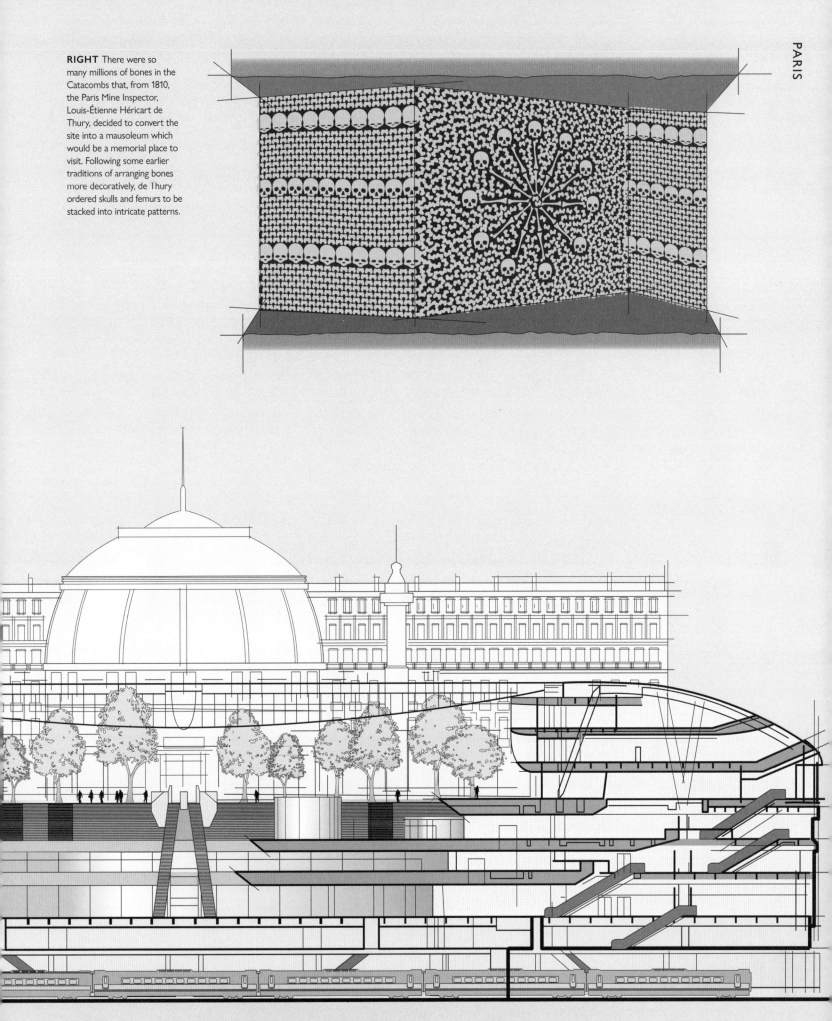

RIGHT There were so many millions of bones in the Catacombs that, from 1810, the Paris Mine Inspector, Louis-Étienne Héricart de Thury, decided to convert the site into a mausoleum which would be a memorial place to visit. Following some earlier traditions of arranging bones more decoratively, de Thury ordered skulls and femurs to be stacked into intricate patterns.

include Cité and Saint-Michel, both constructed in steel caissons that were sunk underground from the surface. The unusual construction can be seen in the entrances, which are built inside the wide steel tubes.

Serving the suburbs

The original Paris Métro lines were all built within, or up to, the edges of the old Paris walls, but many suburbs existed outside them and most wanted a slice of Métro action. During the 1920s and 1930s plans to expand the system beyond city limits began to be realized. Almost every line saw a two or three station encroachment into the near suburbs. Although the Second World War froze developments and not much happened straight after, the 1960s were awash with plans, and an old idea returned for mainline services to pass beneath Paris from all sides of the city.

The Réseau Express Régional (RER) required much new, and deep, tunnelling and a huge interchange station between north–south and east–west lines at the heart of the city. Taking advantage of the relocation of the anceint Paris markets at Les Halles in 1971, the buildings were demolished and a gargantuan hole, known as the Grand Trou, was hewn out at Châtelet for the arrival of the new RER lines from all the cardinal points. When opened in 1977 it became the world's largest underground station, serving at first RER A (east–west) and RER B (running south), which went from there to connect to the old Ligne de Sceaux via new tunnels to its old terminus at Jardin du Luxembourg. The northerly section to Gare du Nord was not opened until 1982.

The RER station at Auber (Line A) was almost as vast. Owing to the high water table, it needed to be built like a concrete submarine 'floating' in water-logged soil. Another east–west RER line (Line C) was needed on the Left Bank to connect the mainline station at d'Orsay (now occupied by a museum) with Austerlitz and this was largely made possible by hiding the line inside 700m (2,300ft) of old embankments along the Seine. Line C opened in 1981. The RER expanded exponentially in the suburbs, but it was not until 1999 that a new 2-km (1.2-mile) tunnelled section opened inside Paris itself. RER Line E connected Saint-Lazare with Magenta (serving both Gare de l'Est and Gare du Nord) and suburban Chelles. A new 8-km (5-mile) section of RER E is now under construction between Gare Saint-Lazare and La Défense and due to open in the 2020s.

During the 1990s, plans advanced for the city's first driverless Métro line: MÉTro Est-Ouest Rapide, or Meteor. It would relieve RER A and overcrowded Line 1, making new trajectories at either end. The first section was ready in 1998 and opened as Line 14. Passengers were amazed by its speed, modernity and platform-edge doors, not to mention the lack of a driver. They could sit for the first time in the very front of the train staring down the tunnel as it sped along.

As Greater Paris grows, so does its need for even better transport. The latest plans include 200km (125 miles) of new Métro network on four lines. The Grand Paris Express project will also see significant extensions of existing Lines 11 (by 5km/3 miles to Rosny-sous-Bois) and 14 (by one stop north to Point-Cardinet, but also to take

RIGHT The central courtyard of the Bibliothèque nationale de France is filled with vegetation, giving staff and visitors the crafty impression that they are looking into a garden... despite the lowest levels reaching a dozen stories below ground.

BELOW Once seen as a forbidding underground place, the 2018 redesign of the Forum des Halles by architects Patrick Berger and Jacques Anziutti allows a much more natural green luminance to filter further down into the complex maze of shops and transit below ground. Named La Canopée, the roof structure was inspired by looking through forest tree tops.

over the Saint-Denis Pleyel branch of Line 13; it will also extend at the other end to head south by 12km/7.5 miles to Orly airport).

Due to open between 2020 and 2030, Line 15 will be a brand new, 75-km (47-mile), all-underground orbital line. All but one of the thirty-six stations, will interchange with another Métro line, RER, suburban rail or tram line and it will all be in tunnel. Lines 16 and 17 will both be 25km (15.5 miles) long, and the latter will serve Charles de Gaulle airport. Line 18 will be 50km (31 miles) long and serve Orly airport.

Second World War activity

With the gathering of dark clouds over Europe during the Second World War, Paris prepared for air raids. The Métro station Place des Fêtes proved a good candidate for shelter, since its platforms lie more than 22m (72ft) below the surface and are some of the deepest in the city. The station was being rebuilt for the arrival of Line 11 in 1935, so the chance was taken to provide an entrance that could be secured against attack. Constructed in a loosely art deco style, the new entrance is bombproof, which was doubly useful, as the station was also used as an underground factory to make aircraft spare parts. Nearby Buttes-Chaumont station served as a deep-level military operations bunker. At least a handful of Métro stations in each arrondissement were used as air-raid shelters including Bolivar, Strasbourg-Saint-Denis and Croix Rouge. A hospital was built inside the Métro passageways at Bastille. In 1956 Line 11 was converted to an experimental type of rolling stock to cut down on noise and vibrations: trains were fitted with rubber tyres (*pneus*) running on concrete tracks. The French-designed system was so successful other lines were switched to use it and the technology exported abroad.

At one end of platform B in Gare de l'Est are some odd-looking barriers around narrow stairs that seem to go nowhere. Below, they open into a collection of rooms occupying around 120m² (1,300ft²) and with a ceiling made of concrete that's 3m (10ft) thick before even meeting the earth above. It is claimed the space was built a few years before the outbreak of the Second World War, to serve as a left-luggage store, but why make the roof so impenetrable? At the time, there was paranoia in Paris that this strategic rail station would be among several that could be targeted for gas attack, so the space was built as a sealable shelter for around seventy railway workers. Ironically the bunker was requisitioned by the Nazis after occupation in 1940 and urban explorers have discovered signs written in German still painted on its walls.

Hidden depths

The Bibliothèque nationale de France was established in 1461 and is now housed in four purpose-built glass towers that resemble upstanding open books on old railway land at Tolbiac. Opened in 1996, the reading rooms and vaults descend by a dozen stories and house an automated document-retrieval system – crucial when there are forty million objects to access.

While the Louvre has almost 40,000 artefacts on display, below the surface are hidden vaults and workshops housing up to eleven times more items. In fact, there are so many other crypts, hidden passages and underground spaces old and new, that Paris rightly earns the name the Gruyère of Europe . . . but one last bizarre space worth mentioning lies beneath the famous Palais Garnier, home to the Paris opera: a man-made lake. It was said to have been formed in 1860, when the builders creating the foundations were so frustrated at the constant water ingress into their pit, that they simply walled it in and constructed the theatre above it. The 3m-deep (10ft) closed pond is even said to contain a phantom fish.

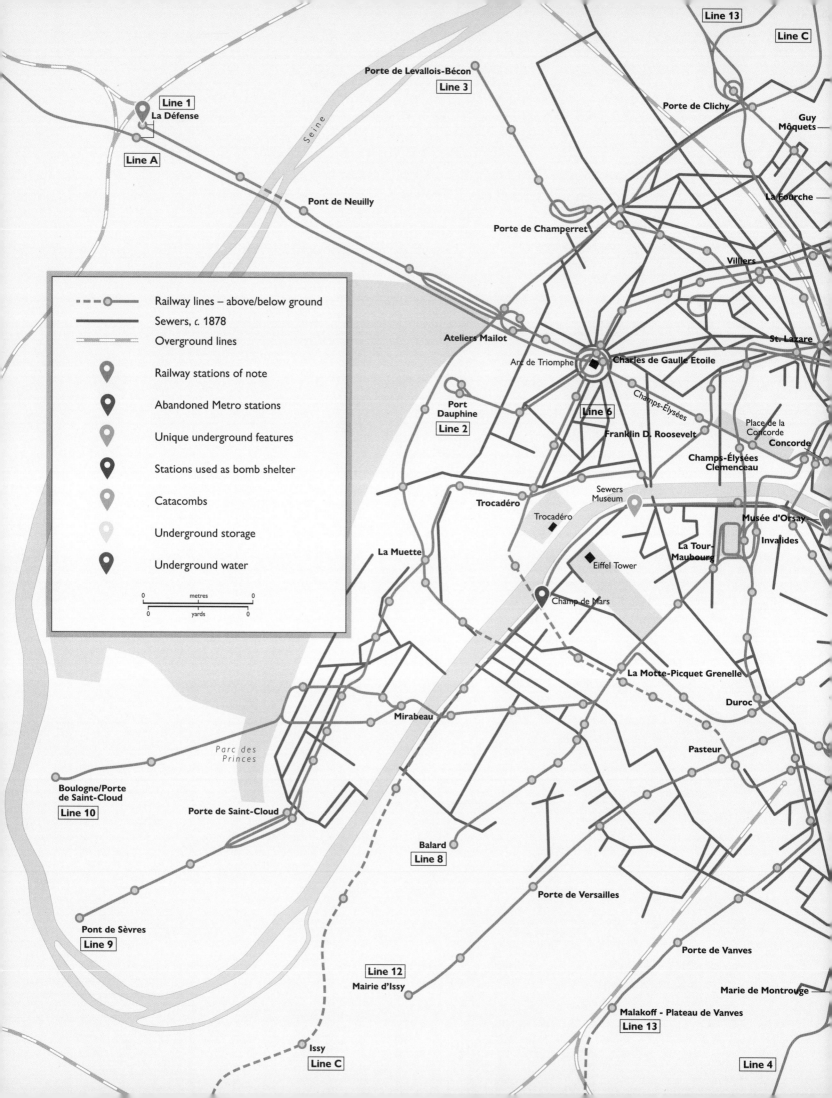

Rotterdam

Holding back the sea

Europe's largest port on the Netherlands' North Sea coast, Rotterdam has 635,000 inhabitants with around 2.5 million in the metropolitan area. It is situated on either side of the wide Nieuwe Mass River, one of several distributor branches of the Rhine delta. Much of the city is protected by dykes as it is substantially lower than sea level.

A town began to take shape here when a dam was built across the Rotte River in 1270. Within a few generations devastating floods prompted the building of new defences against inundation. After just one century of existence Rotterdam was granted city status by the Count of Holland, Willem IV. The Schie waterway linking Rotterdam to other coastal towns, as well as to inland Delft, was completed in 1350. Rotterdam's port activity increased as a result and the city's status was elevated to that of a *kamer* – one of the six regional boards of the Dutch East India Company – during the early 1600s. The Nieuwe Waterweg, 20km (12.5 miles) long, was opened in 1872. An artificial and more navigable canal from the mouth of the Rhine to the North Sea, it further cemented Rotterdam's position as a major port.

The Nieuwe Mass River

Given the depth of the city below sea level – up to 6m (19.5ft) in places – Rotterdam does not have the richest history of subterranean spaces, but one of its earliest major underground constructions was the 1.3-km (0.8-mile) Masstunnel beneath the Nieuwe Maas River. Owing to the river's width, and the volume of shipping using Rotterdam's docks, it was impractical to construct a bridge over the waterway, so work on a combined road and pedestrian tunnel beneath it was started instead, in 1937.

Despite the Netherlands being neutral at the start of the Second World War, Rotterdam was badly bombed and the country invaded by Germany in 1940. Nevertheless, the tunnel project was still completed two years later. Each of its nine 25-m (82-ft) sections was constructed on dry land, floated on the water and sunk into a channel dredged into the bed of the waterway. It was the Netherlands' first underwater road tunnel and the first immersed rectangular tunnel built in this way. It continues to provide a key link in Rotterdam's road, cycle and pedestrian network.

The proposal for trams to pass through a tunnel of their own beneath the Nieuwe Maas was first mooted in 1954, but financial constraints halted progress. Plans for a more substantial metro-type system were approved in 1959. Inspired by the success of building the Masstunnel, a similar method of construction was employed: tunnel sections were built on dry land in three huge construction yards and taken to be 'sunk' into concrete-lined trenches along the city's main thoroughfares, among them Coolsingel and Weena. As a result of the high water table it was an expensive and problematic engineering project. The first 6-km (3.7-mile) stretch of the Rotterdamse Metro, the Noord–Zuidlijn (North–South Line), opened in 1968 between Rotterdam Centraal and Zuidplein.

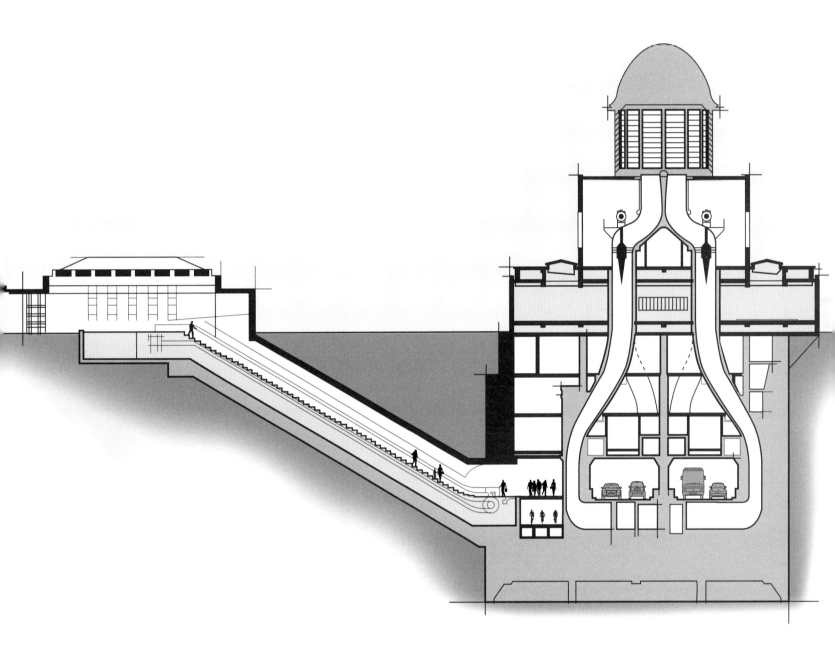

ABOVE At around 20m (65.6ft) below sea level at its deepest point, the Maastunnel was hailed as an engineering triumph when it opened during the Second World War. It was also the first under river tunnel to provide a separate carriageway for bicycles – which is accessed at either end by special escalators reserved only for bikes!

It took a while longer for the second line to come to fruition, but eventually the Oost–Westlijn (East–West Line) between Coolhaven and Capelsebrug was ready in 1982. With the addition of a few new tunnelled sections and other surface lines, the system now connects directly with other cities including Hoek van Holland, De Akkers and Den Haag, running on five routes with one-quarter of the stations in tunnel.

When engineering technology advanced to the stage at which bridges could be constructed over wider waterways, one of the earliest proposals was for a very stylish crossing over the Nieuwe Maas River. Built in 1996, Erasmusbrug (sometimes called the Swan) is an 800m-long (2,600-ft) structure that is both bascule (movable) and cable-stayed. When the bridge opens, a giant counterbalance weighing more than 1,000 tonnes is dropped into a concrete

LEFT Building the first Metro line in the Netherlands took over seven years – and when it opened in 1968 it was then one of the world's shortest lines. As with anything below the water table in this country, construction was extremely expensive and time consuming. This 1961 photograph shows construction of the tunnelled sections on dry land.

BELOW Wilhelminaplein station is unusual in that the platforms gently slope owing to its proximity to the tunnel mouth.

OPPOSITE The storm-surge barrier Maeslantkering is one of the world's largest movable structures. Despite its cost, it was absolutely vital to protect what was then the biggest port on the planet. The barrier will only close when a storm surge of more than 3m (9.8ft) is predicted, which, thankfully, has not happened since the barrier's opening in 1997.

cavern below the road surface, causing the carriageway – which includes a tram line – to tip gracefully up into the air so shipping can pass beneath it.

Subterranean remains

Rotterdam has the rare and dubious honour of possessing an underground ghost tram stop. After the opening of the Rotterdamse Metro, all other transport services were rearranged leaving just one heritage tram line in the south of the city. It was such a busy route that a tunnel was built for it below the Groene Hilledijk in 1969. It even had its own stop, Randweg, but the tunnel was closed in 1996 leaving the abandoned shell as a ghost station.

Underneath the old Post Office, close to the Stadhuis (City Hall) on Coolsingel, is a deep-level shelter that was built by the PTT (the Dutch telecoms company) during the Cold War. Opened in 1975, it was constructed at great expense to ensure the continuation of telecommunications for vital officials in the case of a nuclear or chemical attack. Called NCOs there is a network of these shelters across the Netherlands and the centre is still maintained in case of an emergency.

Storm protection

Given the low-lying nature of much of the Netherlands, and the tendency of the North Sea to inundate it, a monumental storm surge barrier was created in 1997. The Maeslantkering consists of two vast semicircular gates (each 22m/72ft high and 210m/689ft long) that normally sit on dry land either side of the entrance to the Nieuwe Waterweg. The gates can be swung across the ship canal at high tides, where they fill with water and sink into grooves on the bed of the canal to prevent storm surges from reaching the centre of Rotterdam. Effectively holding back the pressure of the entire ocean, the foundations and sunken grooves in the bed of the canal needed to be the most effective ever built. Although the barrier has rarely been used, it is tested every year and has once protected the coast from flooding, although it is anticipated that climate change could see the barrier being needed more frequently than originally anticipated when planned in the 1980s.

The Time Stairs

A large, new, landmark shopping market (Markthal) opened in 2014. The unique horse-shoe-shaped structure rises eleven stories from the surface to house offices and apartments. The market stalls and shops are on ground level, but the basement holds a huge car park that was excavated to four stories below the surface. During the construction, which was on top of a buried fourteenth-century village, many historic artefacts were unearthed up to 7m (23ft) below the current ground level and these have been retrieved and displayed in a museum called De Tijdtrap (The Time Stairs). Visitors are literally walking back in time as they descend back to their cars after shopping.

Amsterdam

Hidden under canals

With an urban population of almost one million – more than two million in the wider metropolitan area – and infamous for its canals and low-lying location, the Dutch capital Amsterdam boasts a few surprising underground features. Until recently, human habitation in this area was considered a relatively new event. Despite its inland location, now almost 30km (19 miles) from the coast, it was a small fishing village in the twelfth century and granted city status in the early fourteenth century. But excavations for the Amsterdam Metro, dug to a depth of almost 30m (100ft) below the surface, uncovered evidence of Neolithic tools and artefacts (hammers, axes, pottery), meaning human settlement was almost 14,000 years earlier.

An abundance of water

Owing to the pressing need to manage water, the Dutch have been damming, bridging and corralling it – and charging fees to cross it – for centuries. The canals were a natural extension of this, so by the 1500s, Amsterdam already had a plethora of moats and waterways either side of the long, thin islands of land on which the city was developing. As it was profiting from its trading routes, a major expansion was proposed in the 1600s, which saw the creation of the belt of semicircular canals radiating out from the centre that characterize the city to this day.

By the seventeenth century and the establishment of the world's first multinational trading corporation, the Dutch East India Company (founded in 1602), the city was entering a Golden Age. The population reached 200,000 around 1660.

Despite the Dutch being water-engineering masters, an obstacle existed between the centre of the old city and its rapidly growing suburbs to the north: the IJ. Formerly a bay, and now more of a river thanks to the region's ever-shifting geological landscape, the IJ proved difficult to cross. In the late 1950s construction began on a new underwater tunnel for vehicles. Opened in 1968, the 1.7-km (1-mile) road is covered for just over 1km (0.6 miles), mainly by the IJ. The deepest point of the tunnel – actually an immersed tube constructed on dry land and dropped into a specially dredged gulley on the riverbed – lies 20.3m (66.6ft) below sea level.

Towards rapid transit

Like most major cities, Amsterdam held aspirations to build a rapid transit system, and the idea of an underground route was discussed in the 1920s. Given that the majority of the city is raised on wooden posts driven into water-logged soil, such a system was deemed technically unfeasible for several decades. A four-line network was finally approved in the late 1960s. Construction work began in 1970 and the first line opened seven years later. It ran 3km (1.9 miles) underground from Centraal Station, before surfacing at Amstel to continue above ground to the suburbs.

Unfortunately for the grand project, cost overruns and protests against the destruction of

ABOVE The long awaited north-south line (opened as Metro Line 52) has some spectacular design features, but one of the most striking is the Centraal Station box and platforms designed by Benthem Crouwel Architects. The firm claims that the shapes below ground were deliberately made to look in keeping with the city: 'mirroring the canals and streets that traverse it at surface level.'

the Nieuwmarkt neighbourhood led to construction of the three remaining lines being cancelled – though not before the interchange for the proposed east–west line had been built below Weesperplein station. This lower level, later converted to a Cold War nuclear bomb shelter, has never been opened to the public and remains one of Amsterdam's secret subterranean spaces, although all bunker equipment has since been removed from the now empty void.

Cancelling something owing to a public outcry, no matter how well meaning, doesn't negate its necessity and eventually Amsterdam's chronic traffic congestion forced planners to act. The proposal to construct a new tunnel linking the north of the city to the south was raised again in 1996. It was rejected by a referendum the following year, but since the result was not binding, test bores were conducted between 1997 and 2002. The ground proved so difficult that expensive soil-injection techniques were required to make safe the 4km (2.5 miles) of tunnelling beneath the heart of the historic city.

Burrowing beneath the canals and fragile historic buildings, the Noord/Zuidlijn needed to be at least 20m (66ft) below the surface. Construction began in 2003 and it was expected that the full 9.2-km (5.7-mile) line would be ready by 2011. That proved to be an underestimate: the opening date was postponed eight times, and Metro Line 52, as it is officially now known, did not enter service until July 2018, cutting journey times across Amsterdam and creating impressive new underground spaces in the watery city.

Subterranean surprises

Despite the fact that a cellar beneath any Amsterdam building would inevitably come close to, or even be forced below, the water table, they do exist. Beside the Nieuwe Kerk (New Church), on Dam Square, is a well-known wine cellar, stocking many South American labels. Entry is via Melly's Cookie Bar and down a spiral staircase.

In the buttress of a bridge in Vondelpark is the entrance to the Vondelbunker. This is one of many bomb shelters dating from the Cold War era. In an emergency, it could have provided safety for around 2,600 people. Functioning as a community centre today, it also houses a microbrewery producing the aptly named Bunkerbier.

Metres	
0	
	Canals (average)
-10	Albert Cuyp car park
	Amsterdam Centraal Station (Metro Lines 51, 52, 53, 54)
-20	IJ tunnel
	De Pijp station (line 52; deepest Metro station)
-30	Metro tunnels (deepest point)

Albert Cuyp Car Park and Weesperplein Station

It is difficult to overestimate the engineering challenges for building watertight structures below sea level in boggy and sodden ground. But the Dutch have demonstrated that determination and innovation are key. So, despite the water table, there are underground facilities like the Albert Cuyp car park, up to 10m (33ft) down; the IJ tunnel; and vast concrete boxes like the one that houses Weesperplein station. Opened in 1977, the platforms sit atop a vast space which has (thankfully) never been used for its original purpose: as a nuclear bombproof shelter.

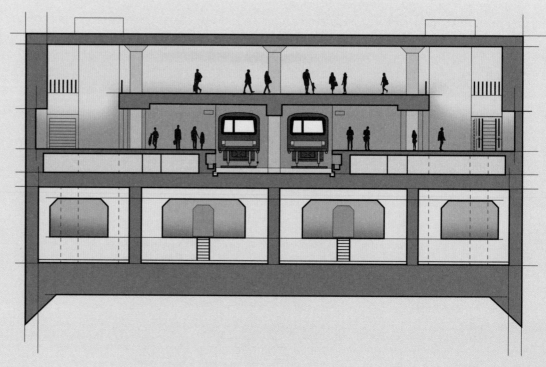

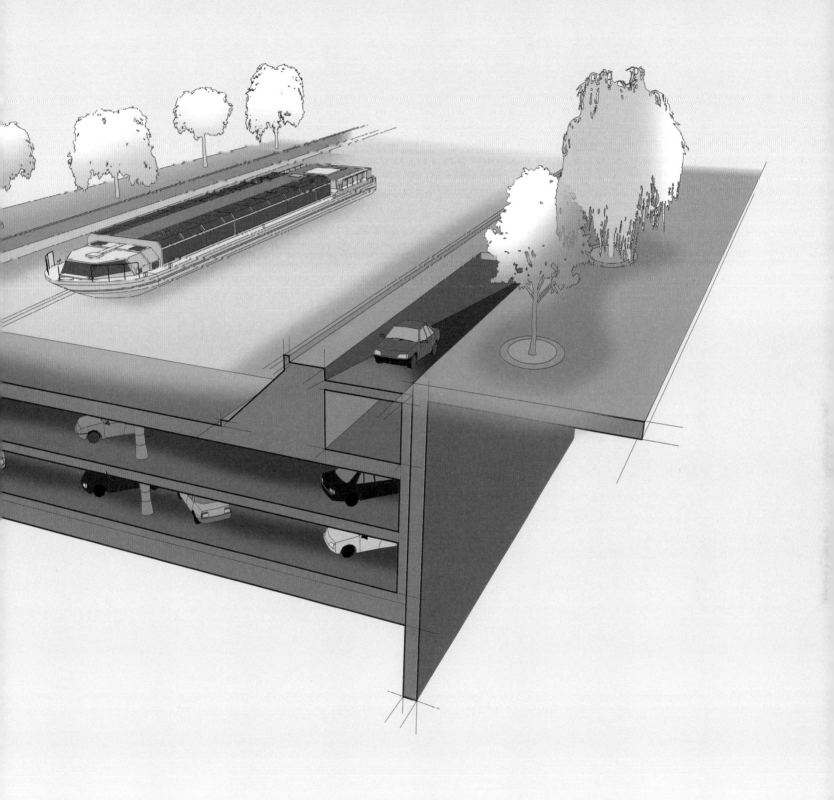

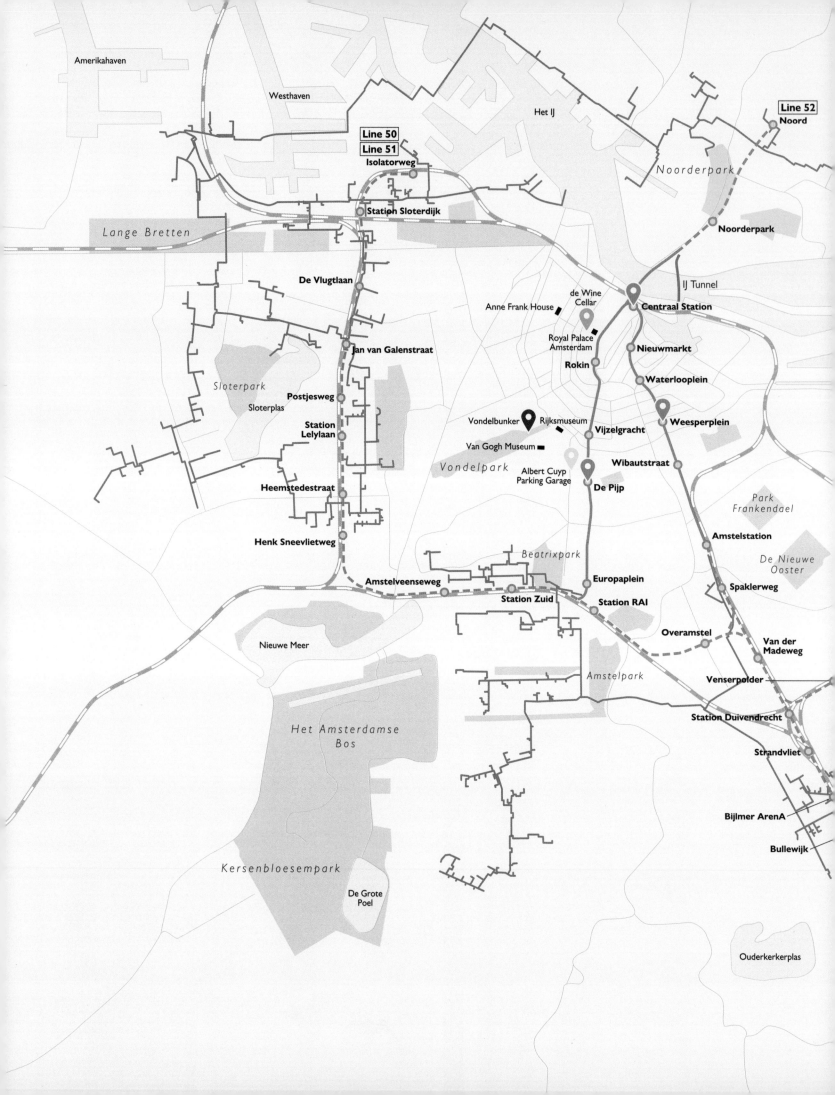

Marseille

Tunnels made a beach

Situated on the Mediterranean coast, and jostling with Lyon as the second city of France, the compact metropolis of Marseille is home to almost 900,000 people, with 1.5 million in the metropolitan area. Founded as a colony of ancient Greece in around 600 BCE, the city had strong connections to Rome and held a pivotal position as a trading port with North Africa. The city's population stood at 100,000 by the start of the nineteenth century.

Canal de Marseille au Rhône
A proposal to link the massive Rhône River directly to Marseille' port was made by the Chamber of Commerce in 1879 and preparatory work began in 1906. The biggest challenge was the stretch of the canal from the large body of inland water, the Étang de Berre to the northeast of the city. In order to bring it close to L'Estaque in the 16th district of Marseille, it needed to pass through the tough rock of the Massifs de la Nerthe to the Côte Bleue on the Golfe du Lion. Once this was achieved, however, canal traffic would be within easy reach of the docks.

The solution turned out to be the longest canal tunnel in the world – a title it still holds. Tunnel du Rove is 7.1km (4.4 miles) long, 22m (72ft) wide and 11.4m (37.4ft) high. Although the tunnel was finished by 1916, it took almost another decade for the full 81km (50 miles) of the Canal de Marseille au Rhône to be fully functional and connect Arles to Marseille. There are so many waterways at Martigues, through which the canal passes, that the area has become known as the Venice of Provence. Although the canal was a considerable short cut for freight, many goods had already been switched to rail conveyance, so the full benefits of the new waterway never came to fruition. The Tunnel du Rove partially collapsed in 1963, causing the closure of the entire canal.

Towards a metro system
Marseille once had an extensive street-running tramway system, opened in 1876 and electrified in 1899. At its peak there were more than 430 cars running on 71 lines. In 1943 came a plan to bury some of the busiest tram routes in two new tunnels through the city centre, but local politicians favoured the car as a mode of transport and the trams were subsequently abandoned by 1960. One route remained open, however: Ligne 68 began operations in 1893 from an underground terminal at Noailles. The tunnel surfaces about 600m (1,970ft) from the stop and runs about 2.5km (1.5 miles) to a terminus at Saint Pierre.

When Marseille was planning a modern light-rail tramway, it was decided to incorporate the ancient survivor. Ligne 68 was closed for refurbishment in December 2003 and integrated into the new tram line T1, which runs mostly on the surface to a new terminus at Les Caillols. Two more surface tram lines have since been added to the network.

An early proposal for a totally tunnelled transport system was made by the Compagnie d'électricité de Marseille in 1918, but never

progressed. Tunnel projects were rejected three times during the 1940s – one plan suggested two tunnels linking Joliette to Chartreux via Saint-Charles station, another (the Monnet plan) proposed three underground tram lines.

It was not until 1964, however, that a more rounded concept for a metro system was given serious consideration as a result of traffic becoming a major issue for Marseille. The Régie Autonome des Transports (RATVM) commissioned a study to replace the busiest bus lines. A route of 7.4km (4.6 miles) serving ten stations was proposed. It would have run from Prado to Chartreux via the Saint-Charles mainline station. Two years later this was fine-tuned into a two-line network of 25km (15.5 miles) serving twenty-six stations. The first (to be called the Blue Line) would run from La Blancard to La Rose, the second (the Red Line) would run from Arenc to Mazargues. The project was approved in 1969, but the state declined the funds, instead organizing a study for how best to serve both Marseille and Lyon.

The delay infuriated locals, but money was eventually forthcoming and construction started in 1973. The total length was revised to 21.8km (13.5 miles), with 18km (11 miles) of it to be in tunnel, and it would run on the French rubber-tyred (*pneu*) system developed in Paris. The first section finally opened in 1977, and the second a decade later.

An unusual by-product of the network's construction was a beach. Despite having a Mediterranean shoreline stretching almost 60km (37 miles), there was no discernable beach in Marseille, so the Prado seaside park was reclaimed by depositing the tunnelling spoil of the first Metro line there in 1975.

Road tunnels

Given the hilly terrain around the city, it was inevitable that, as the road network grew during the twentieth century, so too would the need to tunnel under some of it. Construction work started in 1964 on the 600m (656yd) Tunnel du Vieux-Port and it opened in 1967. Another much longer one (2.4km/1.5 miles) is called the Tunnel Prado Carénage, which opened in 1993. Other road tunnels in Marseille include Tunnel Louis-Rège (300m/985ft, opened in 2007) and Tunnel Prado Sud (1.5km/1 mile, opened in 2013).

ABOVE Guarding the Old Port since 1660, the Fort Saint-Jean can be seen to the right of the Tunnel du Vieux-Port entrance near the Esplanade de la Tourette, and Avenue Vaudoyer.

Milan

Crypts and pieces

The world's design capital, Milan has an urban population of 1.3 million in a metropolitan area that totals 3 to 5 million. Situated in a part of northern Italy that has been inhabited for almost three thousand years, this city has accrued a rich variety of subterranean delights over the centuries.

Milan can trace its history back to around 600 BCE, when the Celts settled the area. It was conquered by the ancient Romans in 222 BCE, becoming the capital of their Western Empire from the year 286. Many Roman monuments remain. A chequered history of sieges and feuds from that time to the end of the Middle Ages kept the rulers and citizens busy, and the city was eventually surrounded by walls in the sixteenth century. Much of what is now northern Italy was under Spanish or Austrian control until the locals rebelled in 1848 and the area joined with Sardinia. Later, in 1861, when Italy was unified, Milan's industrial expansion and railway connections to the rest of Europe cemented its domination of the north.

Historic remains

In the heart of the city centre, about 4m (13ft) below the Piazza del Duomo, is the San Giovanni alle Fonti Baptistery, built in the fourth century. This, the first known octagonal font, was unearthed during the construction of the Milan metro system, the Metropolitana di Milano, in the 1960s. There is also a surviving Roman crypt beneath the ruins of San Giovanni in Conca, at Piazza Missori. At times in its past, the crypt has served as both a mausoleum and a warehouse. There are scores of passages under the famous landmark of Castello Sforzesco. An especially long example is called *la strada coperta della Ghirlanda* (the covered road of the Ghirlanda), which was thought to have been an escape tunnel during a siege, but turns out to have been made to feed water directly into the moat. Another tunnel leads to the church of San Marco, possibly built in the early sixteenth century when Leonardo da Vinci worked in the castle. Some of the castle tunnels may also have been used as air-raid shelters during the Second World War, alongside many tunnels below the city that were specifically built for that purpose. At least twenty of them are being restored.

Milan's trams

Horse-drawn trams arrived on the streets in 1876, serving a route between Milan and the neighbouring city of Monza. Steam trams were trialled just two years later and an entire, intense, inner-city network spread out from the Piazza del Duomo from around 1881. In 1892 the Edison company began electrifying some tram lines. They began numbering the routes (Lines 1 to 30) in 1910. All the local trams that ran through the centre of Milan were taken under the control of the city council seven years later. In the 1930s, hundreds of series 1500 vehicles were purchased and painted in the company yellow. There was also a series of what are called 'inter-urban' lines from Milan to towns further away (Limbiate was

ABOVE Although San Giovanni in Conca Basilica can trace its roots to the fourth century it was rebuilt several times in the eleventh, thirteenth and nineteenth centuries, and partly demolished in 1949. However, the basement survived and is the only remaining example of a Romanesque one in Milan. Some of the frontage at surface level has been restored and it is through here that access is gained to Cripta di San Giovanni in Conca.

the first, in 1882, and Desio soon after). Both systems reached their peak in around 1939.

The 1953 Piano Regolatore Generale Comunale proposed the elimination of all trams, and the commencement of work on an underground rail network – first discussed before the Second World War. But, unlike many other cities and despite the arrival of some trolleybuses, Milan did not give up on the tram. Many of them are still running today, although the system now runs to just 182km (113 miles), comprising seventeen urban lines and one inter-urban. A former tram depot has been converted into an underground spa: QC Termemilano is set beneath the old tracks by the brick arches at Porta Romana.

Metropolitano di Milano

Concepts for mass transit in Milan can be traced back to 1857, when the engineer Carlo Mira proposed a radical plan to divert the water of the Martesana Canal, and then lower its bed, roof it over and lay rail tracks so horses could pull trams along it. In 1905, engineer Baldassarre Borioli had an idea for the mainline railway featuring a circular line about 9km (5.5 miles) from the city centre, equipped with eight interchange stations coming from underground lines radiating out of Piazza del Duomo. That same year Carlo Castiglioni and Leopoldo Candiani proposed an underground line to connect Porta Vittoria with Porta Magenta (also passing through Piazza del Duomo).

In 1912 the city called for further suggestions and received three strong proposals: architect Carlo Broggi suggsted a line that linked San Cristoforo and Loreto; engineer Franco Minorini envisaged burying the busiest tram routes below street level; and electrical engineer Evaristo Stefini proposed a subway link between Milan and Monza. Less than a year later, the city council decided that, even with the backing of big contractors such as Edison, AEG and Italian and French consortia, not one of the three projects met the needs of the city. The only trajectory they supported was a link between the two mainline stations – Central and North – passing through the Piazza del Duomo.

The outbreak of the First World War stopped the study, but Edison came back in 1923 with a pledge to build the line as the council wanted in lieu of them being allowed to take back control of the trams. The councillors rejected this, but instead inserted the idea of a subway into their regional plan of 1933. By 1938 a large network of seven underground lines was proposed, cut to five lines a few years later, but again war stopped any progress.

The council approved a project for four lines in 1952 and work finally began to build the first underground line in 1957. It took seven years before the first trains ran on the initial section of the Metropolitana di Milano between Lotto and Sesto Marelli, via Duomo. A second line, which took over part of a former inter-urban tram route to Gessate, was ready by 1969, but the third took longer to materialize and it was not opened until 1990. Line 5 was ready by 2013, but Line 4 is still under construction. There are now 106 stations running over 97km (60 miles) of track, mostly underground.

With subterranean transit well established, plans evolved to place some of the key suburban train services in tunnels as well. The objective was to connect the northern and western commuter routes to the eastern and southeastern lines via a

ABOVE The concourse between platforms at the busy Passante ferroviario di Milano station first opened in 1997. It is more of a suburban railway with underground sections like this but the full length of the system was not achieved until 2008.

TOP RIGHT The former barber shop Albergo Diurno Venezia was the last business to leave the abandoned bath house complex beneath Piazza Oberdan. Along with the rest, it is currently the subject of a restoration and preservation project and could be open to visitors again soon.

BOTTOM RIGHT One of Milan's most controversial artist's Biancoshock designed this minuscule room inside a drain shaft as part of his work to highlight the exorbitant price of accommodation in cities like Milan.

new subterranean cross-city railway called the Passante. Interchanges at Metro Lines M1 (at Porta Venezia), M2 (at Garibaldi) and M3 (at Repubblica) were also improved. The initial section of the Passante railway was opened between Milano Nord Bovisa and Porta Venezia. It has since become a critical central underground spine for Milan's mainline and commuter trains.

Cultural flair

By far the strangest underground spaces beneath Milan are the works of the conceptual artist Biancoshock. In his effort to draw attention to people living in poverty – such as the sewer dwellers of Bucharest and those living in cramped Milanese apartments – he has been filling abandoned drains with domestic room sets.

So far he has made three of the social artworks, including a kitchen and a shower room underneath unused manhole covers in the streets.

Beneath the Piazza Oberdan lies a sub-surface bathhouse. It was opened in 1926 by the local council covering approx 1,200m^2 (1,300ft^2) and also housed a barber shop, manicurist and, oddly, a photographers and a travel agents. Parts of the site were shut during the building of the first Metro line and although the last occupant (the barbers) left in 2006 it is currently being restored to its former glory.

Oslo

Opportune opening

The fjord-side city of Oslo on Norway's west coast is brimming with Norse heritage and has an official population of just under 700,000 with around one million in the wider metropolitan area. The city celebrated its millennium in the year 2000 – although the Norse settlement of Ánslo at the northern end of the Oslofjord was probably founded in the middle of the eleventh century by Norwegian King Harald Hardrada. It became a capital city when King Haakon V moved there in the early 1300s, but lost its status a century later when the country was dominated by the Danes.

Following fourteen serious fires over a short period of time, little of the original timber-built town remained by 1624 and the settlement was resited as Kristiania (in an area now called Kvadraturen), where a grid layout and more substantial buildings made it safer. A university was established in 1811. Three years later Kristiania became the capital of newly independent Kingdom of Norway. Its status demanded nationally significant buildings and financial institutions, expansion into neighbouring areas, the coming of industry and a large growth in population. Railways arrived in 1854. The name was changed to Oslo in 1925.

Oslo's rail network

A 6.2-km (3.9-mile) light-railway route – the Holmenkollen Line – was opened in 1898 from Besserud to Majorstuen on the western side of the city. The terminus at Majorstuen was still a tram ride away from the city centre. In 1921 construction began on a tunnel between Majorstuen and Karl Johans Gate, 2km (1.2 miles) to the east, more in the heart of the city centre. However, a bad landslide near Valkyrie plass led to a major hiatus while Oslo's council deliberated over the new siting of the end terminal.

At around the same time, the council organized a competition to design a light-rail system for the whole city. The winner was announced in 1918: a new *trikken* (tram) would connect on the surface from Majorstuen to the Lilleaker Line, a street tram already being built, and opened in 1919. A tunnel would be constructed in the city centre from Majorstuen to Stortorvet and then surface to run on stilts towards Vaterland on the eastern side of the city. Running north from Stortorvet, another branch would head towards Kjelsås, which would intersect with an orbital line.

Although this scheme was not realized, the proposals made the council look more carefully at the Holmenkollen tunnel. Construction finally began in 1926 on a 300-m (985-ft) tunnelled section from Majorstuen to the new temporary terminus in Studenterlunden Park by the Nationaltheatret (National Theatre).

Two years later, the opening of the full tunnel from Majorstuen to Nationaltheatret – via Valkyrie plass, where an unplanned station was built owing to the cavity caused by the landslide – was hailed as the start of the first underground in the Nordic countries. Although this section was only 1.6km (1 mile) long in total, it would prove crucial to future mass transit developments in Oslo.

As part of the urban expansion between Oslo and its neighbouring communities, a four-branch mass transit system was planned in the 1950s. A new tunnel was built in 1966, between Tøyen and Jernbanetorget, next to Oslo East station, with an intermediate station at Grønland. The intention was that, one day, it would join the older train tunnel at Nationaltheatret. Two existing tram lines were upgraded to metro standard so they could run in the new tunnel, but that meant they used a different technical specification to the existing lines on the other side of the city. It took over a decade to extend the newer tunnel from Jernbanetorget to reach a new station at Sentrum.

In 1983, water ingress caused a shutdown at Sentrum station, which remained closed for four years for repairs. When the station was ready to reopen again – now named Stortinget – it served as the new terminus for metro lines entering the city centre from the east and the west. By 1993 it was possible for trains to run right through the tunnel from either side of the city, thus creating what is now called the Fellestunnelen (Common Tunnel). Totalling 7.3km (4.5 miles) in length, the tunnel now forms the central spine of the Oslo metro, and is used by all six lines.

The new metro network, which had taken so long to achieve, took the Norwegian name Tunnelbanen (or T-banen). With the opening of the Ring Line in 2006, it now has 85km (53 miles) of track serving 101 stations (17 of them underground), giving the relatively small city of Oslo one of the world's most extensive mass transit systems for the number of inhabitants.

The original section of the Holmenkollen Line, now extended, totals 11.4km (7 miles), but its Valkyrie plass station closed in 1985 because it proved difficult and dangerous to increase platform sizes to accommodate the longer trains that were needed for the conversion of the western lines to metro specifications. This is now the city's only ghost

ABOVE The cross-passageway at Nationaltheatret links the T-banen to the mainline railway station and was part of a project for station improvements in 1995 ready for the start of the Airport Express service (three years later). LPO Arkitektur og Design won with their concept 'Next to Nothing', which is shown here and visible in other parts of the station.

station. The platforms can still be seen in the darkness from passing trains, and it is occasionally used as a location for filming.

The mainline railways also used tunnels under Oslo. A serious landslide in 1953 in the Nordstrand district (a coastal area south of the city centre) blocked the Østfold Line, which had been opened in 1879. A nearby road-widening scheme (the Mosseveien) was combined with the railway repairs, which enabled a 578m (1,896ft) tunnel to be constructed in 1958.

Privately operated rail lines had built stations around the edge of Oslo since rail first arrived in the country during the mid-nineteenth century. They no longer satisfied twentieth-century needs, however, and planners dreamed of linking the city's disparate termini to one another. In 1938 the mainline railways proposed a tunnel between Oslo East (Oslo Ø) and Oslo West (Oslo V) stations. As a result of the cost of overcoming difficult terrain for railway construction, the design was not approved for thirty years and the 3.6-km (2.2-mile) tunnel was not ready to open until 1980.

The initial station at Nationaltheatret (which was aligned directly beneath the T-banen station of the same name) was expanded, at great cost, to accommodate more platforms in 2009. A beautifully designed cross-passageway leading from the mainline station is lined with wood-pressed concrete and brightly coloured stainless-steel sheets.

The Opera Tunnel System

Oslo has an impressive complex of interconnected motorway and road tunnels known collectively as Opera Tunnel. Built in several phases since the year 1990, they extended from Filipstad in the west of Oslo to Ryen in the east. Today, the combined length of this tunnel is about 6km (3.7 miles) with three lanes in each direction, along with various access points. Around 100,000 vehicles per day now pass easily under the city centre, allowing the streets around Rådhusplassen (City Hall Square) to be much less affected by pollution from car fumes.

The first 1.8-km (1.1-mile) section to open was Festning Tunnel, in 1990. This was followed in 1995 by the Ekeberg and Svartdalen tunnels in 2000. Most recently, in 2010, the 1.1-km (0.7-mile) Bjørvika Tunnel – 675m (2,215ft) of it below the sea – joined the system. Also known as motorway route E18, the combined Opera Tunnel system provides much easier access to the waterfront from all directions.

Unconnected to the Opera Tunnel system are two other subterranean vehicle passages: one, called the Granfoss Tunnel which runs from Ullern Church to Mustad (1.5km/1 mile) and the other between Mustad and Lysaker (500m/1,800ft). Both were opened in 1992.

RIGHT The southernmost entrance of the two Granfos Tunnels.

BELOW One of the most expensive but crucial links in the E18 Motorway is the 1,100 m (3,600 ft) Bjørvika twin-bore immersed tunnel in central Oslo. Opened in 2010, with three lanes in each direction, more than half of it is under the sea (the Bjørvika arm of the Oslofjord). It connects Ekeberg Tunnel and the Mosseveien with Akershus Fortress, forming part of the city ring road and the Opera Tunnel network (the interconnected tunnels between Filipstad and Ryen).

Rome

Where roofs become foundations

There are 2.8 million people living within Rome's city limits, and two million more in the surrounding metropolitan area. It sits about 30km (18.5 miles) inland from the western coast of Italy. Uniquely, Rome entirely encircles another country: Vatican City. An independent state, and the world's tiniest, the Vatican lies in the northwestern part of Rome, with a population of less than 1,000.

Ancient history

Archaeological evidence suggests that the area along the banks of the Tiber River was inhabited 14,000 years ago, while the roots of ancient Rome grew from Palatine Hill about 2,800 years ago. Soon every one of the six neighbouring hills was also topped with a settlement, making Rome's one of the longest periods of continual occupation of any European capital.

The Roman Kingdom was founded in 753 BCE and lasted around 250 years. It became a republic during the aggressive territory-grabbing spree leading to the establishment of the great Roman Empire that dominated much of Europe and parts of Asia. The population had risen to around one million by the time of Christ's birth, making it the most powerful city-state in the world at the time. Julius Caesar's son Augustus (the first Roman emperor) and his successors gradually established grandiose buildings such as the Imperial Fora, Trajan's Market, the Circus Maximus and the Forum Romanum. The Colosseum was commissioned by Emperor Vespasian in the early 70s CE but was not finally ready for use until after his death in the early 80s CE.

After the city fell in 476 CE, so did the population, and many of the glorious buildings became ruins. Rome was revived during the Italian Renaissance in around 1400, and underwent a further revival from the 1870s, having become the capital of the newly unified Italian state.

The Cloaca Maxima

It is easy to forget that the low-lying valleys around Rome's hills were effectively below sea level. Most were drained and filled in by hand with rubble, to raise the land level by around 10m (33ft). At the same time, Roman engineers seized the opportunity to allow space for canals, which very soon became sewers. While fresh water was brought to Rome from the neighbouring high land and lakes along magnificent aqueducts, the early sewers were more about draining the marshier land in between the hills than dealing with waste. But the Cloaca Maxima was one of the world's earliest purpose-built sewers. Begun around 600 BCE, it was originally an open canal that was partly covered over during the following decades under the rule of Rome's last king, Tarquinius Superbus.

As growing numbers of latrines and bathhouses were connected to it, more of the Cloaca Maxima was culverted and maps show it running a good kilometre in a southwesterly direction from the Forum Vespasiani to an outfall point into the Tiber River. The portal still survives, though modern sewers prevent raw waste going directly into the river or the waters flowing back up.

A city buried

After the fall of the Western Roman Empire, from the late-fifth century many of the existing buildings were lying empty. When new ones were required, they were simply built up on top of older, decaying ones, often using stone lying about from other ruins. This effectively created foundations and basements beneath the newer constructions, giving rise to many cellars.

Although the same process happened in other cities, it is especially marked in Rome, partly because there were just so many ruins to build over. Given that there were once tens of thousands of apartments and almost two thousand palaces in the ancient city, the vast majority of which had collapsed and been built over, there's a lot of truth in the saying that ancient Rome is literally buried beneath its own remains. Of the many examples, an easy one to access lies beneath the twelfth-century Basilica San Clemente. The first basement level is a hall that still bears faint frescoes of the original San Clemente, one of the city's earliest Christian churches, built 800 years earlier. At some point around 1000 CE, this basement was deemed unstable and filled in with tonnes of rubble and debris, upon which the new basilica was built. Still, it is possible to pass below this level to an even earlier structure that dates from around 100 CE and was once a temple in a private residence. Incredibly even this was built upon an earlier level. Archaeologists have dated this third level to the Great Fire of 64 CE, which raged for six days, destroying two-thirds of the city. It is thought that this level was built over during Emperor Nero's post-blaze reconstruction of the city. Archaeologists estimate the pipes – still carrying diverted streams – and thick man-made brick walls from this, the fourth level, penetrate at least another 5m (16.5ft) down. This pattern is replicated over the entire footprint of the original Roman world. Visitors are quite literally walking over a multilayered, honeycombed history beneath their feet, riddled with spaces that house entire villas, baths, stadia and roads.

BELOW One of the many layers of artefacts beneath the Basilica san Clemente is the second century Mithraic Temple, where pagan followers sacrificed bulls.

Metres

- 0
- Cloaca Maxima
  Necropolis under St Peter's Basilica
- -10
- Roman barracks under Amba Aradam station
- Domus Aurea
  Colosseum's underground chambers
- Catacombs of Domitilla
- -20
- Mussolini bunker beneath Palazzo Venezia
- -30
- San Giovanni Station (Lines A, C; deep Metro station)
- -40
- -50
- -60
- -70
- -80
- -90
- -100

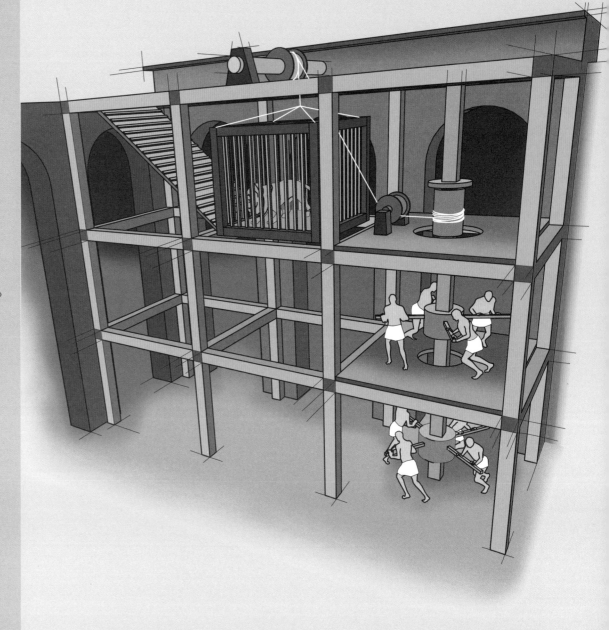

ABOVE The arena of the Colosseum featured a wooden floor (83 x 48m / 272 x 157.5ft) topped with sand. It was supported by hundreds of pillars which created a labyrinthine two level structure known as the hypogeum (a Latin word which translates as 'under ground'). Scenery for the elaborate shows, gladiators, slaves and animals were all kept here for lifting to the surface. Many tunnels lead from the hypogeum to gladiators barracks, prisons and animal stables so that they could be used in the spectacles.

Colosseum and the Domus Aurea

The Colosseum was indeed a colossus: the epicentre of Roman life, completed around 80 CE after about a decade of construction, it is elliptical in shape (189 × 156m/620 × 512ft) and was capable of holding up to 80,000 spectators: the largest free-standing amphitheatre in the world. It was used not just for gladiatorial conquests but acted as a gigantic theatre where huge spectacles were staged including mock battles, dramas and the baiting/chasing of animals (9,000 of which were slaughtered in the inaugural games in 80 or 81 CE). The thoughtful construction of the seating and the inner workings beneath the Colosseum gave rise to its success and popularity as a venue until into the sixth century.

The Domus Aurea was Emperor Nero's vast palace and grounds. Beneath the surface is an example of where the Romans utilized concrete to form two dining areas around an octagonal court topped by a huge dome which had an oculus at the centre to allow light down.

BELOW Artist's illustration of how the light might have poured into the lower areas of the Domus Aurea.

Cities of the dead

Another remnant of Nero's rebuilding lies below Esquiline Hill, where precarious steps on its eastern edge lead down to the Domus Aurea (Golden House), which must once have been a vast opulent palace. A discovery made in 2019 unearthed a room painted with majestic felines that has been dubbed the Sphinx Room by archaeologists. After Nero's death, part of his palace was obliterated by the Colosseum built over the top.

On Capitoline Hill the Senatorial Palace temple is built astride the site of a much older religious building, under which there is evidence of an ancient necropolis. Across the Tiber, and lying more than 5m (16.5ft) below St Peters Basilica in Vatican City is another, larger necropolis which was only unearthed during the 1940s.

A second-century-CE necropolis also exists beneath the slopes of the hill upon which the Vatican now stands. The apostle Peter was interned here, and while Peter's tomb was protected, the rest of the site was filled in and forgotten about. Buried for centuries, some of the mausoleums stretch for at least 30m (100ft). Some were found to be marked with Greek lettering, and others showed evidence of at least one thousand funerals having taken place. In recent years further tombs have been

uncarthed and restored and are available to visit.

Like Paris and other ancient cities, Rome has many catacombs. The largest is under the church of San Callisto where up to 20km (12.5 miles) of tunnel house the bones of Christian pilgrims. The inclusion of sixteen popes gives rise to its nickname of 'Little Vatican'. Another large ossuary, Catacombe di Domitilla, lies 16m (52.5ft) beneath the ground at Via delle Sette Chiese, where 15km (9.3 miles) of tunnels are located. There's another 12km (7.5 miles) of catacombs inside an ancient mine below the Ardeatino Quarter at San Sebastiano. There are more than thirty other smaller catacomb sites throughout the city, for example at Sant'Agnese and Santa Priscilla.

Metropolitana di Roma

Given the riches of Rome's underbelly, it is not surprising that the city was rather slow in developing a metro network, the Metropolitana. The first line was planned in the 1930s under Fascist dictator Mussolini. It was due to run between the mainline railway station Termini, and the proposed site of the Universal Exposition, set to take place in 1942. Work began in an area south of the city centre in the late 1930s. The metro system would have run to 11km (6.8 miles), with more than half of it underground, but the exposition was cancelled because of the Second World War. Some of its tunnels were used as air-raid shelters during the conflict.

The project was restarted in 1948, by which time the location of the huge exhibition site had been repurposed as a new suburb named Esposizione Universale Roma (EUR). This first line (later named Line B) was 18km (11 miles) long serving twenty-two stations and was ready by 1955. Rome's second metro line (later named Line A) did not begin operating until 1980. It runs close to the edge of Vatican City and is almost entirely underground. A short branch – B1, 8km (5 miles) long and 7km (4.3 miles) of that in tunnel – opened in 1990. Line C was an amalgam of an old light-rail route and some new tunnelling in the city centre and it opened in 2014 (19km/11.8 miles and twenty-two stations).

Almost every step of Line C's construction has been interrupted by incredible archaeological discoveries. An entire station site that would have been beside the ancient Forum, with entrances from the Piazza Venezia, had to be relocated a few metres from its planned position following the discovery of priceless historic remains. Named the Auditoria, the ruins are thought to be the remants of Rome's first university. Excavations for Amba Aradam station revealed a long military barracks. In almost pristine condition, it contained forty rooms, some with large mosaics still intact.

So much has been found that the city's museums have been turning things away. At several stations – San Giovanni, for example – large glass display cases have been built, partly to decorate the subterranean passageways, but also to show off the wealth of the finds. Some extensions are still on the drawing board, but the project to build a Line D was shut down in 2012. Rome has 31km (19 miles) of surface-running lines remaining from what was once the country's most extensive tram system. Now modernized, the six routes are serviced mainly by modern light-rail vehicles (with a handful of converted heritage trams that remain in the fleet).

The city also has three commuter railways (Ferrovie Urbane). One of them, the Roma-Viterbo line, terminates in the city centre with a 2-km (1.2-mile) tunnel that actually predates the Metro. It was opened in 1932.

When Metro Line B was extended, it included a station that was built, but laid unopen for thirteen years. Pietralata (between Tiburtina and Monti Tiburtini) was originally constructed to serve a new district called Sistema Direzionale Orientale (SDO), and space was left to accommodate an interchange with the unbuilt Line D. While the rest of the line opened in 1990, Quintiliani, as it was renamed, did not see paying passengers until 2003.

Parking the car

Parking is inevitably a big issue for the modern Romans who don't take public transport. In 2000, when Vatican City needed a new car park, one with 900 spaces was built five stories deep within Gianicolo Hill. Blessed by the Pope, it was dubbed 'Gods Garage'. Another was started three years later, this one directly underneath the Vatican itself. During construction yet another necropolis was discovered – one that was so intact and full of archaeological gems that the parking scheme had to be abandoned. The site is now a museum.

TOP Completed in 19 BCE the Aqua Virgo was one of eleven major aqueducts which effectively transported thousands of cubic litres of fresh drinking water into the city. Stretching 21km (13 miles) into the hills above Rome, it was almost entirely underground and had one of the shallowest gradients of all the aqueducts – gently dropping only 4m (13ft) from its source. This spiral staircase gives access to it near the Villa Medici and the Spanish Steps.

BOTTOM Located in the Complesso Callistano (a 30 hectare funerary network) on the Via Apia, this third century CE fresco on the entrance to the Catacomb of Callixtus depicts Saint Cornelius and Saint Cyprian.

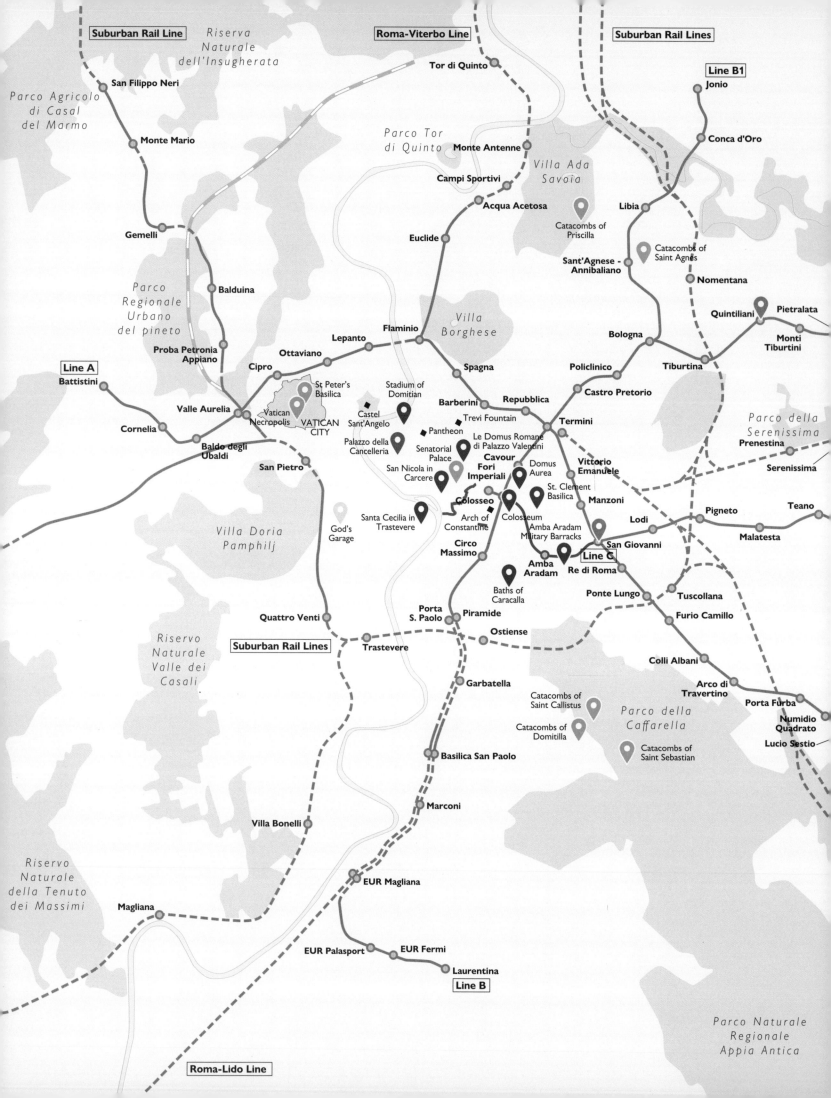

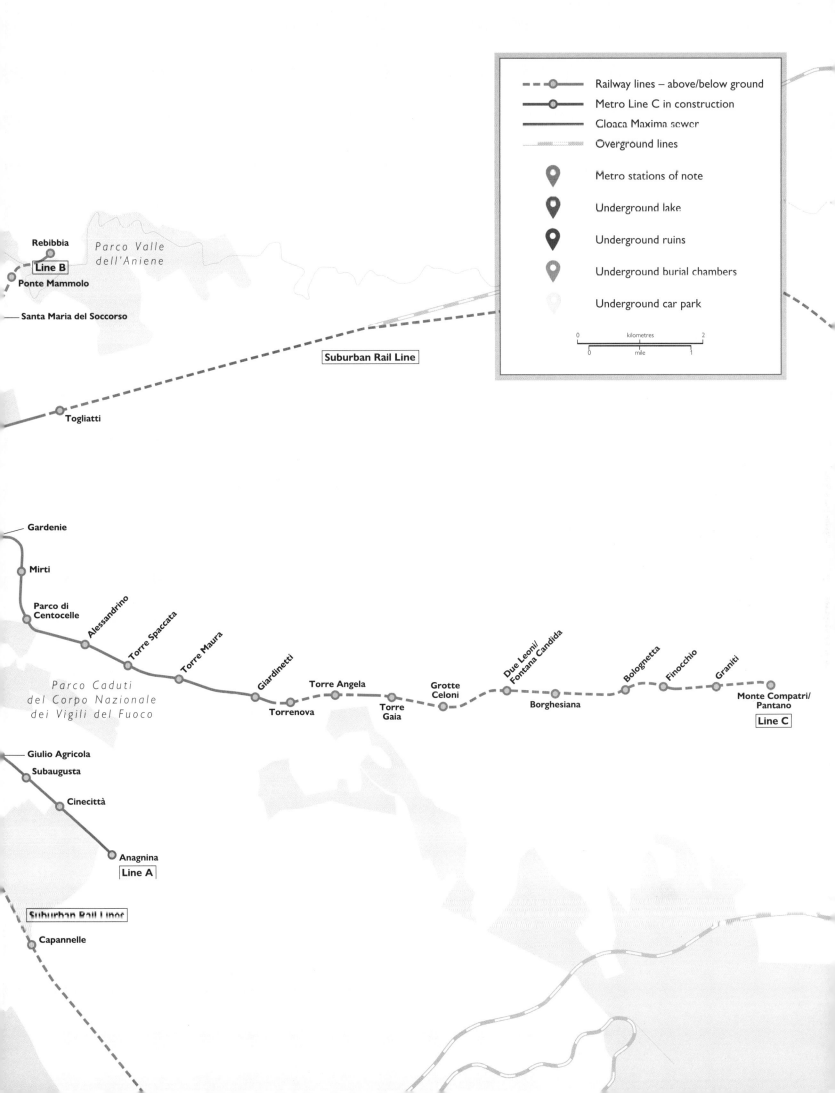

Munich

From the ashes

Home to many museums, Christmas markets and the Oktoberfest (beer festival), Germany's third-largest city and Bavaria's capital has 1.4 million inhabitants with six million people living in the wider metropolitan area.

Founded by Benedictine monks on the Salt Road, an ancient trading route, Munich was first documented in 1158, receiving fortifications and city status in 1175. During the early-sixteenth century, the city became a centre for the Renaissance and came into its own when Bavaria became a sovereign kingdom in 1806.

Many of Munich's grandest neoclassical boulevards and buildings date from the early-nineteenth century and the rule of King Ludwig I. It was also during his reign that Germany's first railway was built, running from Nürnberg to Fürth, in 1835. Named after the king, the Bayerische Ludwigseisenbahn (Bavarian Ludwig Railway), along with the initial 12-km (7.5-mile) stretch of the Munich–Augsburg line, which opened in 1839, helped kick-start the industrialization of the region.

Munich's U-Bahn system

Early proposals to build an underground rail system in Munich arose in 1905. One route would have run between Ostbahnhof and Hauptbahnhof, with another providing a Ringbahn – an orbital line. Both were considered excessive for the traffic of the day, however. Instead, the tram network was felt sufficient to deal with passenger numbers in the then half million population. The metro idea was resuscitated in 1928, with a proposal for a five-line network, but this was quickly thwarted by the global economic crisis. From 1936, Munich was seen as the centre of the Third Reich movement, so it was deemed worthy of investment in an urban rail system. Construction began in 1938 on the first tunnel of an S-Bahn route between Sendlinger Tor and Goetheplatz. By 1941, the 590m (1,935ft) tunnel was complete and the first railcars were to be delivered the same year. Scarcity of resources during the Second World War led to a suspension of work, however. The tunnel was re-purposed as an air-raid shelter and was mostly back-filled with rubble after 1945.

The reconstruction of Germany following the destruction during the war was not a speedy process. Munich had been decimated, and the residents voted to rebuild the old town rather than opt for a Frankfurt-style modern look. Building tops were capped at the height of church spires, American-style shopping malls were not permitted and the city centre looks resolutely historic today.

The cost of such measures was the lack of investment in a mass transit system. A tram network was proposed in the 1950s, but rejected in favour of a full metro system – a U-Bahn – which took longer to deliver. The old 1940s tunnel at Goetheplatz proved sturdy enough to be utilized for one of the running tunnels, so construction began on a parallel tunnel in 1964, to house the first 12-km (7.5-mile) north–south U-Bahn route between and Goetheplatz and Kieferngarten (now Line U6). In 1966 Munich was

TOP Bavaria is famous for its beers, and many breweries close to the Isar river stored thousands of industrial sized barrels of it in brick-lined cellars beneath the city. This artist's impression gives an idea of their scale compared to the barrels delivered to bars and cafes.

BOTTOM It is evident that much thought was put into the decor of many U-Bahn stations. Opened in 1971, the colour theme at Marienplatz is vibrant orange (a fairly common choice for 1970s subways). This later passageway extension was clad using the original colour scheme.

awarded the right to host the 1972 Summer Olympic Games, so a branch was built from this first route, running from Münchner Freiheit to Olympiazentrum (now part Line U3). At the same time, a new east–west tunnel was built straight through the heart of the city to link the mainline rail stations on either side of the central area. This took the name S-Bahn (a heavy-rail line as opposed to the narrow guage and mainly underground U-Bahn) and interchanged with the U-Bahn at Marienplatz.

In 2017 the first earth was cut for second cross-city S-Bahn tunnel. The new 7-km (4.4-mile) route is needed to provide more capacity and faster journey times between Hauptbahnhof and Ostbahnhof, and includes a new station at Marienhof. Dubbed the 'Crossrail' of Munich (in reference to a scheme in London), the €3.84 billion, ten-year build, has taken sixteen years of planning, and is due for completion in 2026.

Water management

Owing to its position on the floodplain of the Isar River and the Bavarian climate, the city required a major strategy to avoid flooding. Flood water used to pick up pollutants, which ended up back in the river, so then tween stormwater retention structures have been constructed since the 1970s. The combined system now has a total storage volume of 706,000m^3 (25,000,000ft^3). Storm water is stored temporarily in the vast underground tanks before being let back into the sewer system in a controlled manner.

Berlin

Divided and healed

Germany's capital is also the country's largest city, with an urban population of 3.7 million and a further 2.3 million living in the wider metropolitan area. What originated as a small, thirteenth-century village surrounded by lakes at the heart of the European Plain, served as the capital of a province known as Margraviate of Brandenburg during the fifteenth century. As a map from 1688 shows, the four earliest settlements that now form the city of Berlin grew effectively as a collection of populated islands in the Spree River, with Berlin and Cölln connected to each other via two main bridges and to the land around their walled fortifications by many others. These settlements subsequently became the centre of the Kingdom of Prussia, established in 1701, and merged with nearby Friedrichswerder, Friedrichstadt, Dorotheenstadt and Cölln eight years later, under the single name of Berlin.

The city grew rapidly during the Industrial Revolution, swallowing up neighbouring settlements and becoming a rail and freight hub. With Prussia having been the driving force behind unification, Berlin was named capital of the new German Empire in 1871. By 1920 the city's surface area had expanded from 66km² to 883km² (25 to 341sq mi) increasing the population to almost four million – Europe's third largest municipality. Divided after 1945, at which time it no longer served as the capital, the city was reunited in 1990 when it was returned to its traditional status as Germany's capital.

Below the surface

City planners had to address the issue of waste water early on in Berlin's development. Given its location on a glacial spillway that was once home to many lakes and the floodplains of rivers such as the Spree, it was impossible to remove accumulating sewage without a concentrated effort and the city would have eventually drowned in its own pollution. Berlin's first waterworks was opened in 1856 at Stralauer Tor, but it wasn't until fifteen years later that a municipal construction company led by James Hobrecht began the huge task of building proper brick-lined sewers. Berlin now has around 9,700km (6,000 miles) of wastewater drains and sewers, about one-third of which are primarily designed to cope with storm water.

Despite the city's long history, very little else seems to have been dug under it until the need for transportation arose in the early 1900s. The Berlin Horse Railway Company began running trams between central Berlin and nearby Charlottenburg in 1864. When the system was electrified, it required overhead wires, which were deemed a little unsightly for the more beautiful boulevards, so a solution was suggested to place electric trams serving Unter den Linden beneath the boulevard, leading to the inauguration of the Lindentunnel in 1916.

Meanwhile other transit systems were already burrowing beneath the surface. An early experiment of a tunnelled rail line was built by the AEG company in 1897. The 300-m (984-ft) tunnel ran some 6.5m (21.3ft) below the surface between

two of the company's Volkspark Humboldthain factories. Electric trains carried both freight and factory workers. The line was also seen as a trial for future underground railways in the city, but the authorities were nervous of disturbing their new sewage system.

Germany was well advanced with its railway development and even by the end of the 1800s, Berlin was almost entirely surrounded by elevated heavy-rail tracks, some of which are still used by today's S-Bahn. The first line of the narrower gauge rail that later became the U-Bahn began in 1902 with the majority of its 10km (6.2 miles) above ground. Only four stations, and 2km (1.2 miles) of track were in tunnel, but it was extended several times between 1908 and 1930. The Nord-Süd-Bahn (North–South Line), for which construction began in 1912 did not open until 1923 because of the war. This was the first of what became known of as 'large profile' lines as a result of the bigger rolling stock used. A rapid expansion of existing lines and new ones continued until the beginning of the 1930s.

Legacy of conflict

In the run-up to the Second World War, the Nazi government began constructing thousands of air-raid shelters. Although hundreds took shape as fortified towers on the surface, many, such as Gesundbrunnen Bunker, were huge subterranean excavations. Today the Berliner Unterwelten e.V, a society dedicated to preserving Berlin's underground sites, provides tours of the bunker and several of the city's 'ghost' U-Bahn stations. A few other bunkers, such as that found at Blochplatz and which could have sheltered more than 1,300 people, have survived, but the majority have been filled in or destroyed.

Nazi leader Adolf Hitler's ostentatious plan to rebuild Berlin to create a new world capital 'Germania' with grandiose buildings and monumental arches and columns, was mostly

ABOVE Gesundbrunnen is the biggest and best preserved air raid shelter from the Second World War. Meticulously maintained by Berliner Unterwelten e.V. the Unterwelten Museum tour provides an insight into bunker life as well as exploring some of the U-Bahn's history, the sewerage systems and the once busy pneumatic postal distribution network. Members who provide the tours claim visitors will see up to 160 years of Berlin's underground history.

Metres

- 0
- Vorbunker
- Bernauer Strasse escape tunnels
- AEG Experimental U-Bahn
- 10 Führerbunker
- Tunnel 57
- Tiergarten tunnels
- Gesundbrunnen Station (Line U8; deep U-Bahn station)
- 20
- 30
- 40
- 50
- 60
- 70 Berlin water wells
- 80
- 90
- 100
- 170

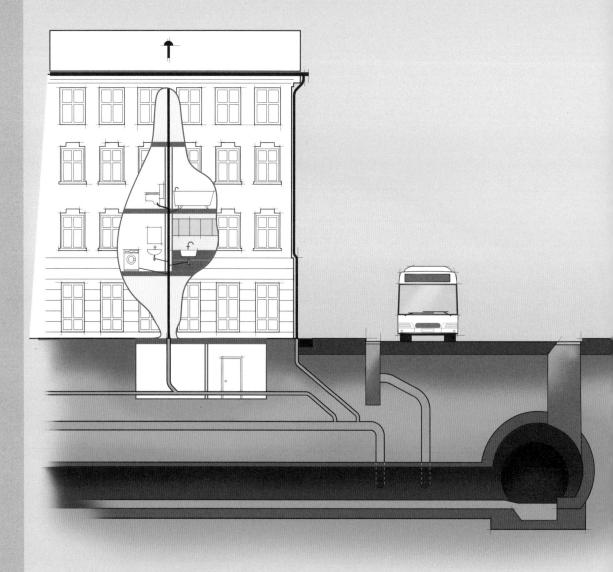

Combined and Separate Sewer Sytems

Deliberately constructed to accommodate all three main types of sewer (stormwater, wastewater and combined water), Berlin's sewerage system totals a length of over 9,700km (6,027 miles). Around three quarters of the city is set up for the collective system (below left), in which wastewater and stormwater are transported together. This design is extremely practical for the city centre in particular, where public transport routes already occupy a lot of space beneath the streets. The rest of the city collects the wastewater and stormwater separately (below right), which has the advantage of coping better with especially heavy rainfall.

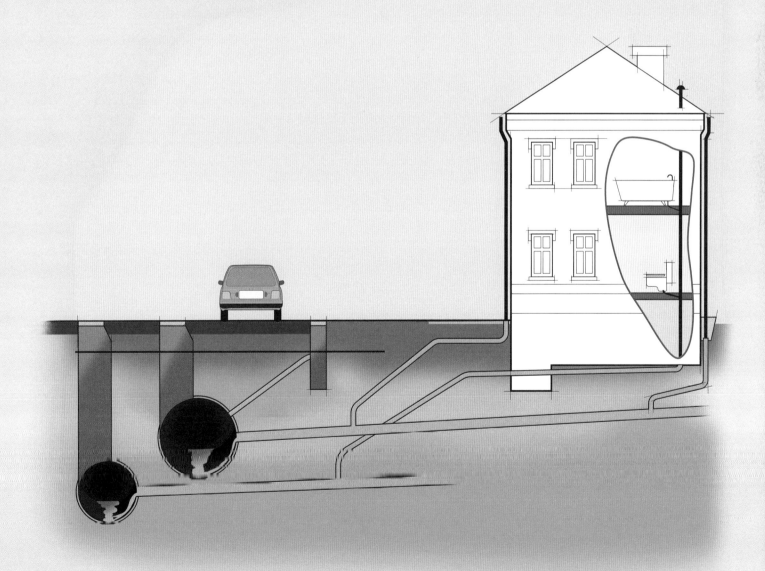

ABOVE The cellars of the former Oswald-Berliner Brewery now serve as exhibition rooms for one of Berliner Unterwelten's tours. Here visitors can learn about both failed and successful escape tunnel attempts from the time of the Berlin Wall.

unfulfilled. Under the direction of architect Albert Speer, the city was to be centred on a vast, new 5-km (3-mile) ceremonial boulevard called Prachtallee, some of which ran along the widened Charlottenburger Chaussee – now Straße des 17. Juni – with another section running along the line of the Siegesallee, a former Kaiser's project, through the Tiergarten, later returned to parkland. This 'avenue of splendours' as it roughly translates, was intended as a parade ground. Closed to traffic, vehicles were instead to run beneath the boulevard. Some tunnelling work for this subterranean highway was carried out and remains in situ.

The Iron Curtain

Following its defeat in the war, Germany was divided between the Allies (the United States, the United Kingdom, France and the former Soviet Union). Since Berlin fell entirely within the Soviet area, that too was divided into four sectors. Although inner borders were erected inside Germany and loosely in Berlin itself, some 3.5 million East Germans (20 per cent of the population) and other East Europeans fleeing Sovietization defected to West Berlin, many via the U-Bahn, which was still connected between the eastern and western halves of the city. In 1961 the East German authorities closed a number of U-Bahn stations and reinforced the border crossings, banning people from travelling between the two. They also began to construct a 140-km (87-mile) wall to completely enclose the western sectors of Berlin. East Berliners continued to make attempts at traversing the barrier, the most successful using the sewers or building tunnels directly beneath them. As many as seventy different tunnels were constructed over the life of the wall's existence between 1961 and 1989, and although many people lost their lives in collapses or capture, well over 300 people were able to defect to the West through them. One of the most infamous routes was via Tunnel 57, built in October 1964.

The Iron Curtain that characterized the division of Germany and Berlin and the ensuing Cold War, also gave rise to the creation of hundreds of military bunkers, control rooms and potential shelters from nuclear attack. Pankstrasse bunker, built as late as 1977 alongside the U-Bahn extension of Line 8, could have housed 3,339 people for over a fortnight in the event of a third world war.

Reunification ramifications

After the Berlin Wall started to come down in 1989 and Germany was reunified in 1990, all the services that had been severed during the division needed reinstating. One of the first consequences of this, just two days after the official fall of the wall on 11 November 1989, was that Jannowitzbrücke U-Bahn station was reopened to the public as an official border crossing point. It was quickly followed by other ghost stations such as Rosenthaler Platz and Bernauer Strasse. When border controls were entirely abolished in 1990, stations that had remained closed for so many years could be used again without any restrictions, enabling East and West Berliners to be united again, legally, through the tunnels under their own city. Over the next few years, other lines such as U2 and the U1 section over Oberbaumbrücke – the latter had been completely severed during the years of division – were rebuilt.

A victim of reunification was the propsed U10 Line, which had been part of East Berlin's 200-km-plan (125 miles). Developed between 1953 and 1955 and still live until at least 1977, the route would have taken lines diagonally across the city from Falkenberg to Alexanderplatz and Steglitz to terminate in Lichterfelde. Proposed interchanges with a number of other lines led to the construction of platforms and cross passageways in at least five existing stations (Rathaus Steglitz, Schloßstraße, Walther-Schreiber-Platz, Innsbrucker Platz and Kleistpark). The route became known of as the 'Phantomlinie' (ghost line). A West Berlin plan for a new part of U3 to serve Weißensee also collapsed, but only after works had been completed at Potsdamer Platz, where abandonded platforms serve as an events space today.

In the Mitte district of the city, a large U-Bahn station designed by Collignon Architektur is under construction at Rotes Rathaus (near the town hall). It will be up to 32m (105ft) deep in places and will form part of the U5 Line. During construction, remains of the first medieval town hall of Berlin were discovered, which will eventually be visible from an 'archaeological window' in the station's design. It is scheduled to open in late 2020.

BELOW Artist's impression of how the new deep-level station being built at Rotes Rathaus will sit beneath its namesake above.

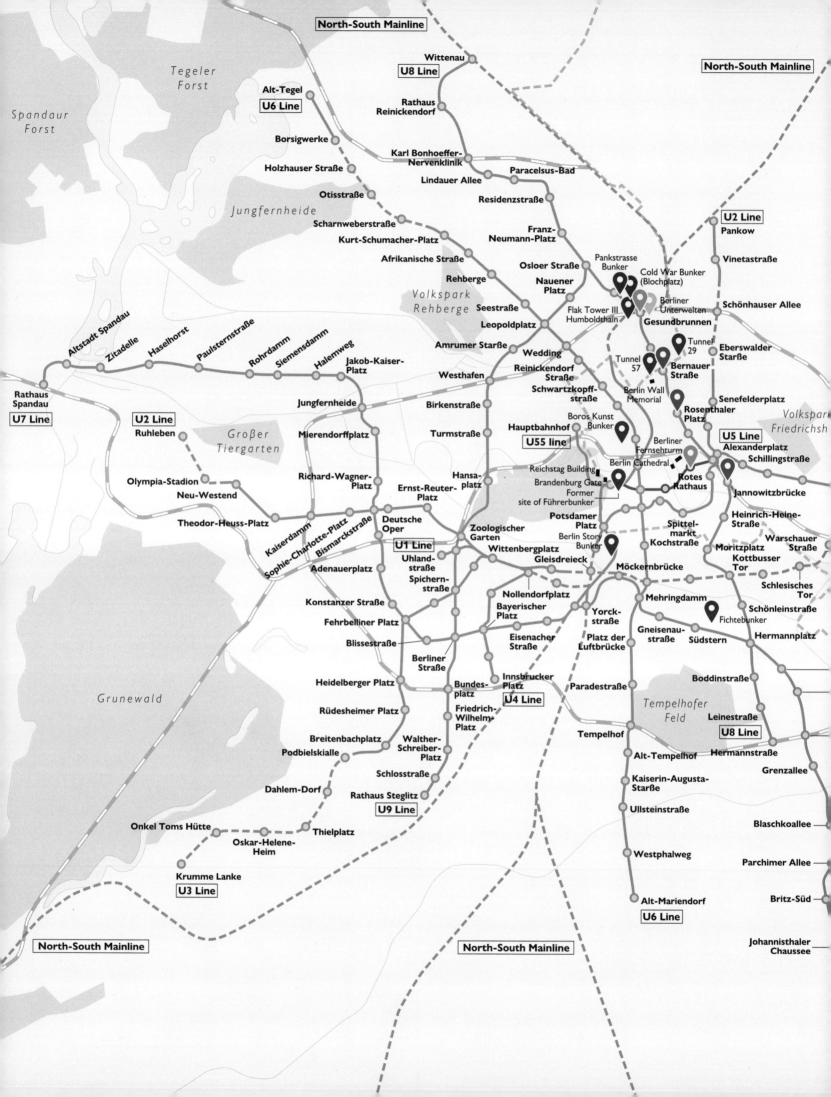

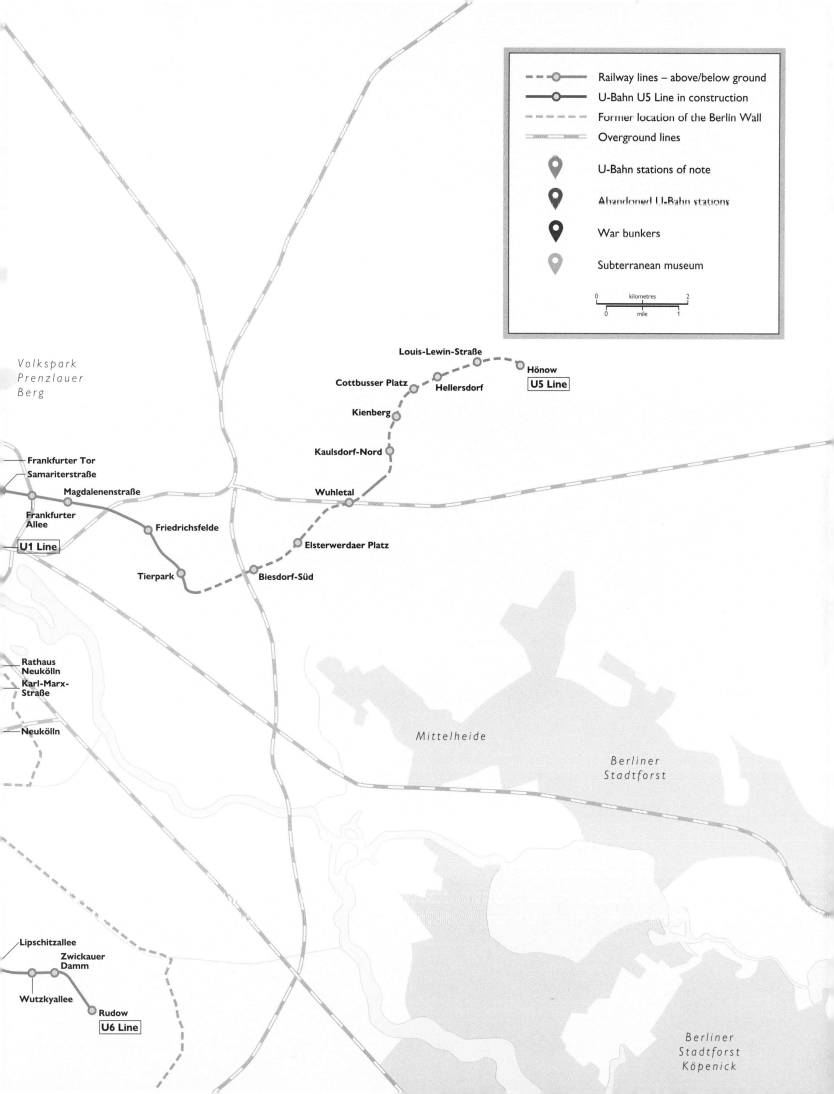

Budapest

Thermal layers

Home to 1.75 million people, with over a million more in the surrounding metropolitan area, the Hungarian capital Budapest began life as a Celtic settlement called Ak-Ink, after which it became known as the ancient Roman town of Aquincum. In the ninth century, the Magyar peoples arrived here from the north of Bulgaria. The ancestors of today's Hungarians, they founded a state under the rule of the region's first king, St. Stephen. The neighbouring cities of Buda, Óbuda and Pest on either side of the Danube became centres of the Renaissance in the fifteenth century and they merged, taking the name Budapest in 1873. The city was also named capital of Hungary and co-capital of the vast Austro–Hungarian Empire at this time (Vienna being the other). Budapest is often cited as one of most beautiful cities in world and is a major East European centre of finance, culture and tourism.

A karst landscape

The original city was built on the site of eighty geothermal springs, which remain the source of the mineral-rich waters used in Budapest's many spas. Beneath the surface, these waters have carved out the largest known grouping of hydrothermal caves in the world, with more than two hundred in total. Although many of the caves are accessible and lie right beneath the streets of Budapest, an even larger number are filled with the hot springs that made them. Divers have been exploring the 70°C (158°F) waters for decades and now well over 6.5km (4 miles) of the network have been mapped.

The caves beneath Buda Castle are mainly dry. Carved out nearly half a million years ago by the hot-water springs cutting through the limestone rocks, lies a vast labyrinth of basements, caverns and cellars. It is thought that prehistoric people used them as refuges hundreds of thousands of years ago. In more recent human history the caves have had various uses. In the Middle Ages they housed torture chambers and prisons, while twentieth-century applications include wine cellars, a military hospital, a secret bunker and a nuclear fall-out shelter during the Cold War.

Gellért Hill, which dominates the Buda side of the city, is linked to this labyrinth, and home to a cave named after Saint Ivan who purportedly healed the sick using muddy sediments taken from the waters. Pauline monks who rediscovered the cave in the 1920s consecrated it and converted it to a chapel. The monks themselves were persecuted by the state, which had them arrested and charged with treason on Easter Monday 1951. The superior monk, Ferenc Vezér, was executed and his fellow devotees forced into hard labour. The chapel was sealed up, but since the fall of the Iron Curtain in 1989, has undergone restoration.

St John's hospital by Buda Castle opened a new hidden wing during the Second World War – in a cave. Named the Sziklakorhaz (Hospital in the Rock), it occupied 2,000m² (21,500ft²), and could house three hundred patients and forty medical staff. In the final days of the war, during the siege of Budapest, it is said that over six hundred patients were squashed in. Connected by passageways to

TOP RIGHT A brave diver explores the thermokarstic limestone Molnár János Caves beneath the city. These are usually only formed in ice-rich permafrost near to the Arctic but there are a few examples like this in the Himalayas and the Alps.

BOTTOM RIGHT Beneath the Buda Castle there are also many caves, and some of the underground spaces like this section beneath the castle walls have been linked by man-made tunnels.

Metres

- 0
- Millennium Underground Railway
- 10
- Castle labyrinth
- 20
- Main sewer under Castle Hill
- Kálvin tér Station (Line M4)
- 30 Kőbánya cellar system
- 40
- 50 Rákosi-bunker
- 60
- 70
- 80
- 90
- 150 Molnár János Cave
- 250

Kálvin tér Station and the Hospital in the Rock

Given Buda and Pest owe much of their history to what lay beneath, it is no surprise that modern Budapest has not been shy about basing much of its infrastructure under the ground. Having built the first subway line on mainland Europe in 1896, it took a long time to supplement it with other routes but much has happened to improve public transport in the last fifty years, including the opening in 2014 of Line 4 which intersects with Line 3 at Kálvin tér.

BELOW A longitudinal view of Kálvin tér Station on the newest Metro line, M4. It also has passive provision for another future line, the purple M5, which itself will link to the city suburban rail network, HEV, making Kálvin tér the best connected place in the city.

BELOW RIGHT A cross-sectional view of Kálvin tér Station, showing the escalators down to the island platform for Line M4.

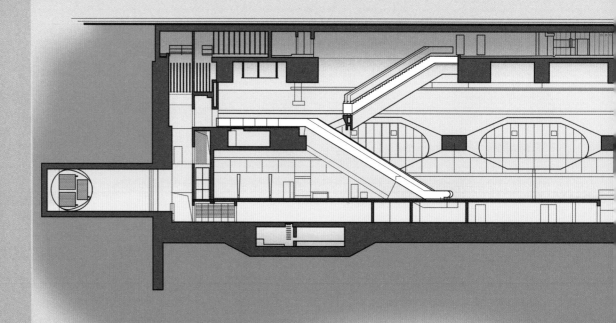

ABOVE With war looming during the 1930s, the city's mayor wanted a secret hospital and air raid shelter built beneath the Buda Castle. It played a major role during the Siege of Budapest (1944–45) and is now restored and known of as The Sziklakórház Atombunker Múzeum (Hospital in the Rock Nuclear Bunker Museum).

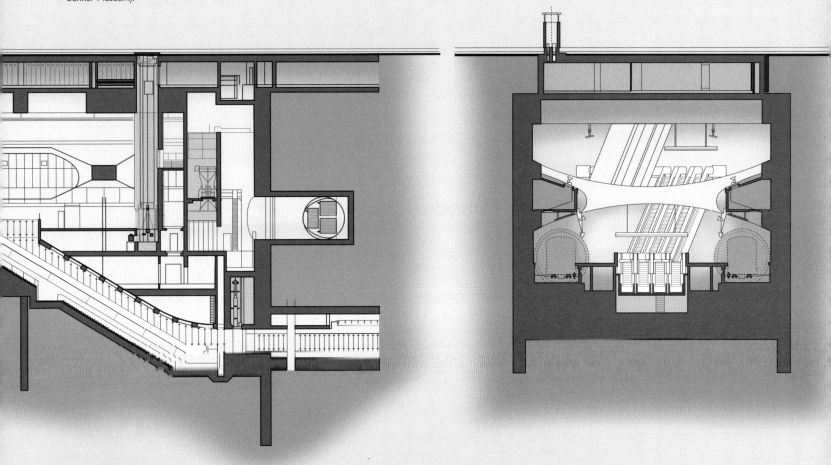

the surface buildings of its neighbouring medical facilities, during the Cold War the space served as a shelter against potential nuclear attack. It was converted to a museum in 2008.

The communist era was tough for the population of Budapest as a whole, but extremely dangerous for dissidents. Troops of the former Soviet Union suppressed the Hungarian uprising of 1956, forcing objectors to hide underground to seek shelter from their attacks. Many are said to have hidden in the caves and tunnels below Budapest – others simply went missing, taken by the authorities and presumed killed. Entrances to some of the escape routes are said to lie around the II. János Pál pápa tér (Pope John Paul II Square, formerly Köztársaság tér/Republic Square) and near to Keleti railway station.

Ahead of the game

Budapest was growing into a grandiose city by the end of the nineteenth century. In 1896 the city was due to play host to a Millennial Exhibition to mark one thousand years since the founding of the Principality of Hungary. Having seen the success of London's underground railway, and with news of other cities wanting to emulate that, the head of the Budapest tramway authority Mór Balázs, argued in favour of building a new tram line to serve the exhibition. Running beneath the boulevard known at the time as Sugár út (now Andrássy út), as the underground route needed to be ready for the big event, the line was called the Millenniumi Földalatti Vasút (MFAV; Millennium Underground Railway), later abbreviated to Földalatti.

Despite passing over a canal and needing to cross a main sewage pipe, it took just twenty months to build the railway, although the tunnel profile of just 6m (19.7ft) wide and 2.75m (9ft) high, was smaller than would be desired now. The 5-km (3.1-mile) route between Vörösmarty tér and Széchenyi fürdő was the first electrically propelled underground line on the European mainland when it opened and was not altered for another eight decades when it was extended to Mexikói út.

ABOVE The deepest station (32m/105ft) on the new Metro line M4 opened in 2014 is Szent Gellért tér – Műegyetem station. The striated mosaic ceilings and walls were designed by Tamás Komoróczky. The ceilings above the escalator shafts also feature an ad hoc looking arrangement of concrete beams.

RIGHT The high altar and congregation seating at the chapel inside Saint Ivan's Cave on Gellert Hill is beautifully lit to highlight the undecorated rock. It is just one of several rooms, another of which features the carved hardwood ornaments of a former Pauline monk.

Any further dreams of expanding the Metro system did not come to fruition until the 1970s. Plans for an east–west line – later known as M2 – go back to the 1940s. Construction began in the 1950s but was halted from 1954–1963. The first new section was not ready until 1970. The north–south line, which became known as M3, was much quicker to materialize. Initial plans were made in 1968 and it was opened by 1976. Much deeper than their older cousin, the Földalatti (now known as M1), at 16.5km (10.25 miles), M3 is the longest of the four existing Budapest Metró lines.

Plans for the fourth line, first hatched in 1972, took much longer to come to fruition. Construction of M4 did not begin until 2004 and it took a further nine to complete the 7.4 km (4.6 mile) route. Future expansion plans are on hold owing to the cost overruns of M4. A fifth route will not be a traditional metro, but a Paris RER-style mainline heavy railway, running through the city centre in tunnels and far out into the hinterland.

Subterranean curiosities

An intriguing relic of the Cold War is Rákosi-bunker, 50m (164ft) beneath Szabadság tér (Liberty Square). It was long rumoured to have been built in the 1950s, as a nuclear fallout shelter for communist leader Mátyás Rákosi and Hungarian Working People's Party apparatchiks. But it was not uncovered until the construction of the M3 Metro line in the 1970s. There was a direct connection from the bunker to the old offices of the party.

In a district once synonymous for mining, a new industry hides below ground. The Kőbánya Mine on the Pest side of the river had created caverns and tunnels stretching over 30km (18.6 miles). Since the end of its life as a working mine, the huge system of tunnels has been purchased by the local authority and a brewery. Besides storing beer barrels and other imperishable products such as canned food, it is the site of a peculiar farm in which thousands of tonnes of mushrooms are cultivated and grown in the old mineshafts, metres below the sunlight.

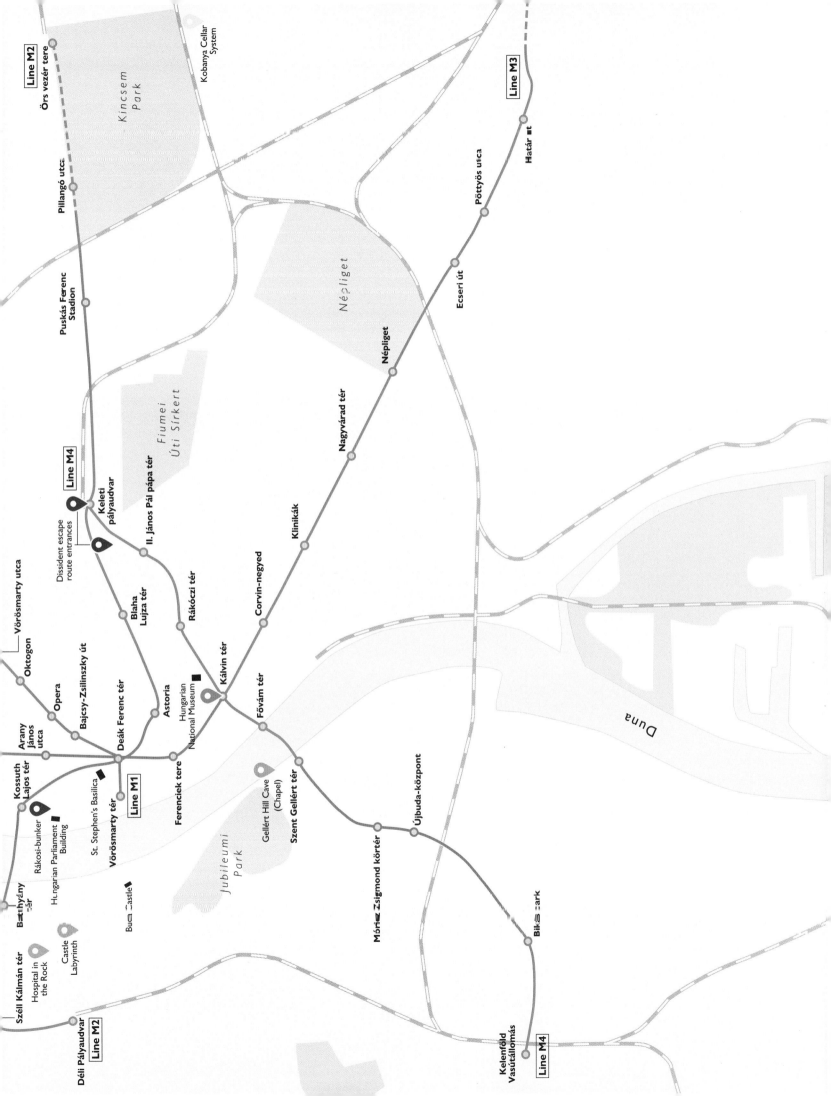

Stockholm

In love with tunnels

The capital of Sweden, the fifth-largest country in Europe, Stockholm is situated on a series of fourteen islands that have multiple connections to the mainland. Almost one million inhabitants are based within the city limits, with 2.5 million in the wider metropolitan area. Although the water around the city looks like sea, it is actually part of the 120km-long (74.5-mile) Lake Mälaren, which leads into the Baltic.

Origins and early infrastructure

While there is evidence of human habitation following the end of the last Ice Age around 8000 BCE, it was not until the 1200s that the current

BELOW After building two other lines (Green and Red), Stockholm Tunnelbana became more ambitious when it opened its deepest line yet (the Blue Line, in 1975). They took the bold decision to leave the exposed rock of the blasted platform caverns undecorated. Although some colour was added via light or paint, the unsculpted look at Rådhuset station shown here has been described as organic architecture (a term first coined by Frank Lloyd Wright in 1954) and, combined with many fascinating and innovative designs for the other stations, provides the backdrop to what has been dubbed the world's longest art gallery.

settlement was established on the tiny island of Stadsholmen, now Stockholm's Gamla Stan (Old Town) district. It is home to many of the city's most historic buildings including the cathedral Storkyrkan, Riddarholm Church (Riddarholmskyrkan), and the stock exchange on Stortorget (the main square). A big trader in the Hanseatic League – a confederation of northwest European towns and merchant guilds – the city grew with strong German-speaking connections, and had a population of 10,000 by the start of the seventeenth century. It was named capital of the Swedish Empire in 1634, after which it suffered a decline resulting from plagues and warfare.

With the arrival of industrialization during the nineteenth century, the city re-established both its trading position and population, topping 100,000 inhabitants in the 1850s and trebling that in just fifty years. The city's sewage system was slow to catch up, however, and was not built until the 1870s. In the 1930s city planners built the world's first wastewater treatment plant inside a hill. From 1936 to 1941, 90,000m^3 (3,200,000ft^3) of rock were removed from inside Henriksdal. After five extensions (the latest in 2015), the plant is now almost double the size of the original. A new 12-km (7.5-mile) wastewater tunnel is also being built to connect the Järva sewage tunnel under Lake Mälaren to Eolshäll pumping station at a depth of 26m (85ft).

City transportation

Street-running trams arrived in 1877, and were electrified around the turn of the twentieth century. Separate lines from the north and the south were connected at Slussen in 1922, and the network rapidly became very popular, with many services sharing the same tracks and causing traffic jams. Congestion was so bad in the southern suburbs that, in 1933, the central section of the line between Slussen and Skanstull on the island of Katarina-Sofia, was placed in a tunnel.

The success of the new services, known as Rapid Trams, encouraged planners to take a decision in 1941 on a future mass transit scheme for the whole of the expanding city. In fact, the Slussen to Skanstull tram tunnel had been built to a high specification in anticipation of it being emulated for a future metro-like service. A new section of converted tram line was built between Skanstull and the southern suburb of Hökarängen.

Metres	
0	
	1933 Slussen to Skanstull tram tunnel
10	
	Henriksdal Wastewater Treatment Plant
20	
	Jarva–Eolshäll wastewater tunnel
30	Barkarby Station (under construction)
	T-Centralen Station (Lines T10, T11, T13, T14, T17, T18, T19)
	Kungsträdgården Station (lines T10, T11; deepest T-banen station)
40	
50	
60	
	Förbifart
70	
80	
90	
100	

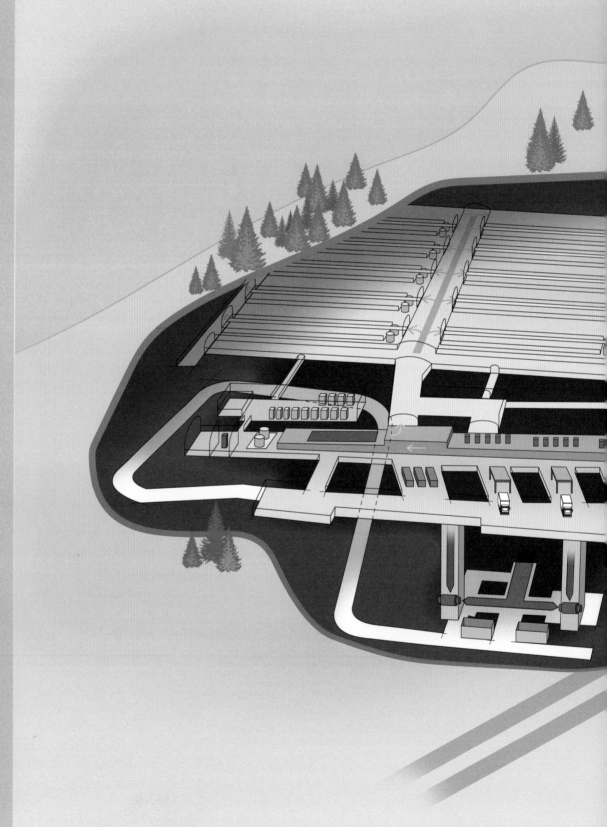

Henriksdal Wastewater Treatment Plant

One of the world's biggest underground wastewater treatment plants has been excavated out of a mountain at Henriksdal, just on the edge of Stockholm. First opened in 1941, it has been regularly expanded to cover a total area of 300,000m³ (186.4 mi³). Using over 18km (5 miles) of tunnels it now deals with the wastewater of over 1 million people every day.

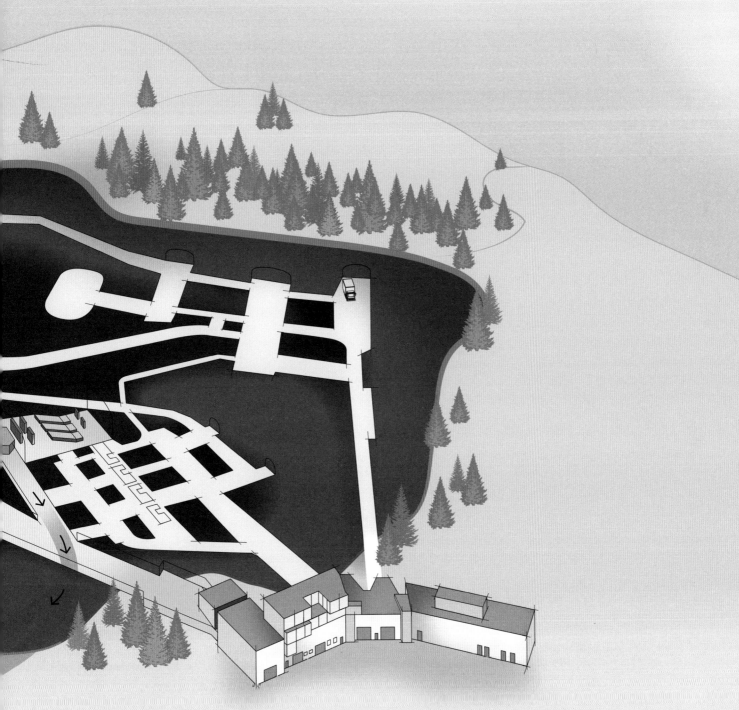

ABOVE An artist's impression of the Blue Line platforms, 32m (105ft) below ground at T-Centralen interchange station. The stylized flower motifs painted by artist Per-Olof Ultvedt are some of the most iconic and recognizable of any metro station in the world – fitting for the hub of the system.

Rolling stock of a higher capacity than the old trams was purchased, allowing metro-style operation between Slussen and Hökarängen from 1950. The line became known as the *tunnelbana* (tunnel rail). A year later a second tram line, from Stureby, was converted to enable it to run into the tunnel at Skanstull. In 1952, an as yet unconnected system from Hötorget, just north of T-Centralen, now the main hub of the Stockholm Tunnelbana, was opened to the western suburbs.

In 1957, a new tunnel was made from Slussen, via a new station at Gamla Stan and under T-Centralen to Hötorget so that the two systems were connected together to form what is now known as the Green Line. The tunnel section between Fridhemsplan and Skanstull became the busiest part of the system and city planners drew up ambitious proposals in the 1960s for many more miles of underground rail services. In 1964, a second, deeper-level line was established. It ran from T-Centralen and split into two branches on the western side of the city: one to Fruängen and the other to Örnsberg. It later headed north towards Mörby and what is now known as the Red Line was essentially complete in 1975.

Owing to the depth of the caverns holding many of the stations on the Stockholm Tunnelbana, they were literally blasted out of the rock. At two underground stations on the Red Line – Alby and Masmo, on the long extension to Norsborg – the exposed rock was deliberately left as it was after blasting. The effect was so striking that when work was underway on the new Blue Line, which was to be almost entirely underground and up to 30m (98ft) deep in places, all stations were left with exposed rock. Each of them was lit or painted differently and between them they have created a unique look for the city. Several are inventively decorated and include Per Olof Ultvedt's stylized blue plants in T-Centralen; the clouds and Pride rainbows at Stadion by Åke Pallarp and Enno Hallek; and the hint of formal French garden statues in red, white and green at Kungsträdgården.

Stockholm's Tunnelbana, which now covers 108km (67 miles) has 100 stations. A total of 62km (38.5 miles) and forty-seven stations are underground. The three coloured lines (Red, Green and Blue) and their branches make seven routes altogether, and they have been dubbed 'the world's longest art gallery', since ninety of the stations are now fully decorated. There is even one ghost station at Kymlinge, where platforms were built half in the open air, but work was halted and they are now abandoned.

The Swedes have taken so well to tunnelling that the central spine of Stockholm's mainline railway network has now been diverted into a new underground section that passes beneath the city centre. This development has provided much needed relief for the formerly cramped surface mainline station, which originally opened in 1871. Since 2017 all commuter routes use the new 7.4-km (4.6-mile) deep-level tunnel, allowing national services more space in the Central Station above.

And the tunnel love is not confined to railways: Sweden is currently building the world's second-longest underground road. Förbifart Stockholm (Stockholm Bypass) has been planned since the mid-1960s. At one stage, a section of the road was to run on a series of bridges across unspoilt countryside, but now the 21-km (13-mile) stretch between Häggvik interchange north of Stockholm and Kungens Kurva south of the city is going to be mostly in tunnel – 17km (10.5 miles) of it. Preparation began in 2009. At its deepest point it will be 65m (213ft) under Lake Mälaren, and when complete it will be second only to Tokyo's Yamate Tunnel in length (see page 200). Construction began in 2014 and is expected to take over a decade to complete.

Mountain room

Buried in caves beneath the Swedish capital lies what was once an impressive military facility. At the start of the Second World War, more than 1,000m^2 (10,800ft^2) of rock were excavated to form the Skeppsholmen Caverns to provide a control room for the navy. The Bergrummet (mountain room) – is now used for occasional exhibitions.

RIGHT Fourteen artists contributed works to the Hötorgen station, which opened in 2017. Pictured here are the jagged fluorescent neon lights in the western hallway by artist David Svensson, who based his 'Life Line' concept on the heartbeats of his son shown on a CTG monitor during childbirth.

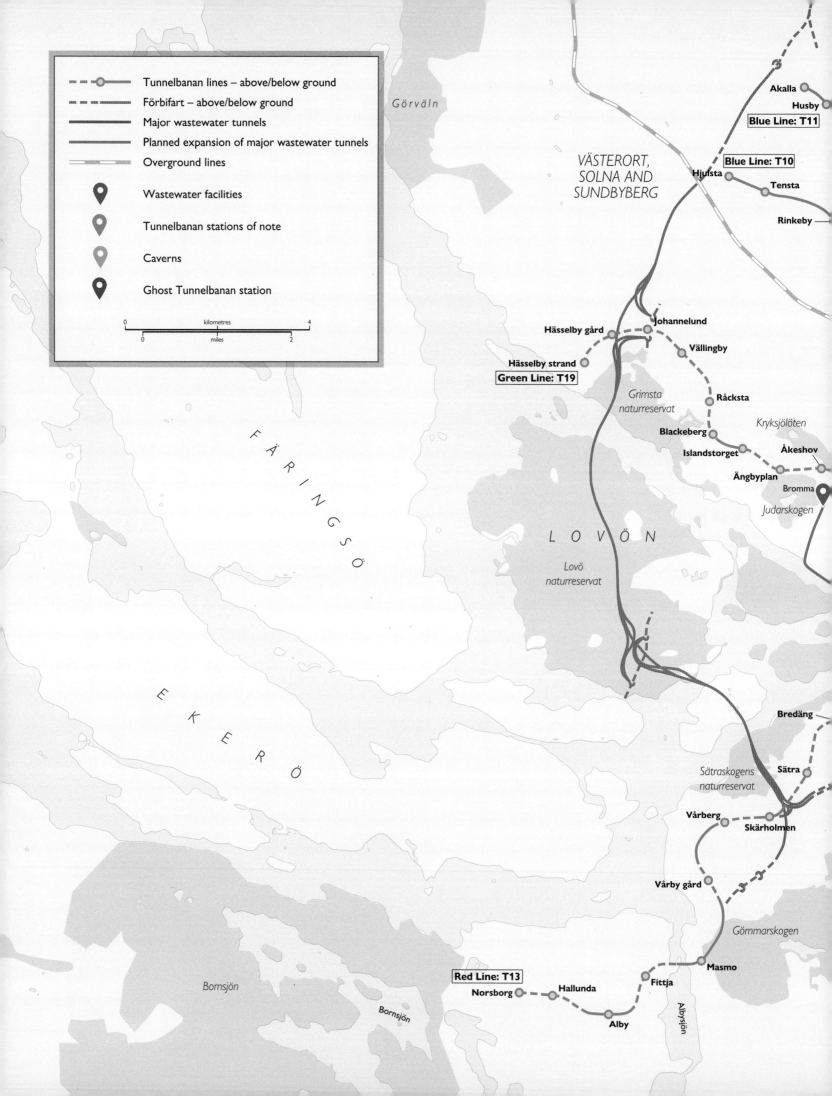

Helsinki

Sheltering an entire city

Europe's northernmost capital, Helsinki has a population of over one million people. The city itself has 648,650 inhabitants and Greater Helsinki, also known as Finland's Capital Region, has a population of 1.4 million. Little was dug underground during the city's early years as it sits on strong bedrock, but the twentieth century changed that.

Early history
The region in which Helsinki is now situated was occupied as far back as the Iron Age. In the 1300s, Helsinge village was established at the tip of a peninsular surrounded by more than three hundred islands. It became a trading town called Helsingfors in 1550, but despite grand plans to rival Tallinn in present-day Estonia, it remained small and poor. The town started to grow during the Grand Duchy of Finland era, after 1809, and became the capital in 1812. It has been known as Helsinki since at least 1819.

A university was established in the city in 1827, and the central area was rebuilt by the German architect Carl Ludvig Engel. His love of the neoclassical style has bequeathed the city a centre that is in some ways similar to St Petersburg in Russia. The focal point at Senate Square, flanked by Helsinki Cathedral and Government Palace, led to the nickname 'The White City of the North'. Industrialization brought railways, and Finnish architect Eliel Saarinen's art nouveau Central Station is a masterpiece of the period's style. In comparison to many other European capitals, Helsinki was a late developer. But winning the right to stage the 1952 Summer Olympics was a boost to the entire country.

Cold War anxiety
Owing to Finland's strategic position in Europe and, in particular, its 1,340-km (833-mile) border with Russia, during the Cold War the Finns were (and remain) wary of an invasion or attack from the other side of the 'Iron Curtain'. Over the decades, provision has been made for an incredible 600,000 residents to be housed in underground shelters beneath Helsinki. Literally hundreds of kilometres of tunnels have been excavated and more than 9,000,000m^3 (11,800,000ft^3) of rock have been removed to create around four hundred shelters, many of them interlinked.

Although the majority of the shelters were created at the height of the Cold War in the 1960s, they are still regularly maintained and kept up to date with the latest technology. There are frequent drills and exercises that involve the military and some civilians, and parts of the vast complex offer more than rudimentary shelter. There is a full-size ice hockey rink, as well as several swimming pools and enough supplies to keep an entire city safely fed and watered for some weeks in the event of an attack.

Helsinki's metro system
The city's metro system was first conceived in 1955. At the time, although it was realized tunnels would be needed, the kind of vehicles that should run in them remained undecided. A report was presented seven years later that proposed an 86-km (53-mile) light-rail system with tunnels running about 14m (46ft) below the surface and serving more than one hundred stations.

The proposal prompted planners to allow room for it on new bridges and a void for a station was created in 1964, when a large shopping centre at Munkkivuori was extended. Although the first plan was deemed a little ambitious, there was a growing realization that Helsinki was approaching a size that could do with at least some kind of rapid transit/metro service.

TOP This tunnel was constructed as part of Helsinki Energy's switch to bio fuel processing. It is 30–60m (98–197ft) deep and runs for 12km (7.5 miles).

BOTTOM One of hundreds of entrances to the bomb shelter network under the city. They are usually marked on the doors with blue triangles, the international symbol for civil defence.

A scaled-down proposal in 1969 envisaged three lines (some heavy rail) and a 2.8-km (1.7-mile) test track was completed by 1971. The first metro line ran between Puotila and Kamppi and most of the tunnelling under central Helsinki had been completed by 1976.

All the first metro stations were somewhat 'overengineered', as they were designed to dovetail into the bomb shelters, so providing extra capacity for the bunker system. The first section of the line opened in 1982. Several extensions have been added since that time and the system now has twenty-five stations, seventeen of which are below ground.

There are several extension projects in the pipeline. Plans for a westerly route towards Kivenlahti are advanced and the line should open in 2022. A new line, nicknamed the Pasila metro, would start at Kamppi and would then run north to Pasila for four stations. In the distant future, it may also reach the airport and then run south towards Santahamina island for at least seven stations. There could be other extensions to Vlikki and Majvik too, but these are longer term projects.

Master plan

In 2010 city planners adopted the extensive Helsinki Underground Master Plan for the development of all the transport proposals and civil defences. Its aim is to provide continued expansion and maintenance of the system so that it is ready for occupation in the event of an emergency, but also so that parts of it can be utilized at any time as a resource for the community above ground. There are even ideas for it to include a subterranean Guggenheim museum.

Metres
- 0
- Typical Finnish cellar
- -10 Itäkeskus swimming hall
- -20
- Bedrock civil defence shelters
- Metro tunnels (average)
- -30 Pasila reservoir
  Kaampi Station (Lines M1, M2; deepest Metro station)
- -40 Lake Päijänne water tunnel (average)
- -50
- -60
- -70
- -80 Maintenance tunnels
- -90
- -100 Esplanadi Park reservoir

Itäkeskus Swimming Hall and Kamppi Station

Under the 1958 Civil Defense act, all Finns are entitled to protection in war and peacetime. The Interior Ministry must make shelters available to everyone who lives in areas at high risk of being attacked – which is most urban areas. It is thought that over three and half million people, mainly in the cities, could survive the detonation of large nuclear weapon in the now enormous network of underground shelters. The Helsinki Metro system was deliberately constructed to act as both a transport system and a formidable shelter against nuclear attack. In the eastern suburb of Itäkeskus, the shelter contains the largest underground public swimming hall.

BELOW The Itäkeskus swimming hall can hold a thousand bathers and even has two water chutes and a 5-m (16.4-ft) high diving board. Opened in 1993 there are actually four pools and a fitness centre – all carved out of the rock and buried from sight.

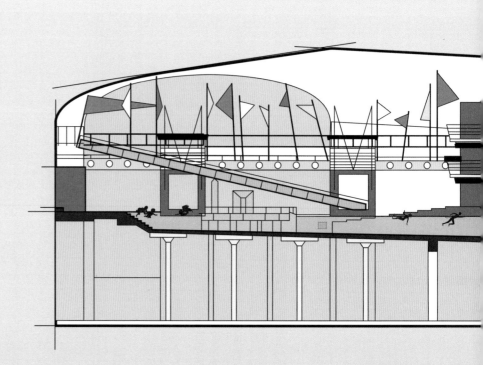

RIGHT The Metro station at Kamppi in the central area was tunnelled alongside existing civil defence shelters and now provides extended capacity to the huge Kamppi-Forum complex, a largely underground shopping mall. The station platforms are the deepest on the Metro at 31m (101.7ft), although beneath those are other platforms intended for use for a future line to be built perpendicular to the alignment of the existing tracks. The clusters of signs on the station's ceiling were commissioned by Helsinki Art Museum. Celebrating Heslinki's multiculturality, they point to some of the varied places Helsinki residents originate from.

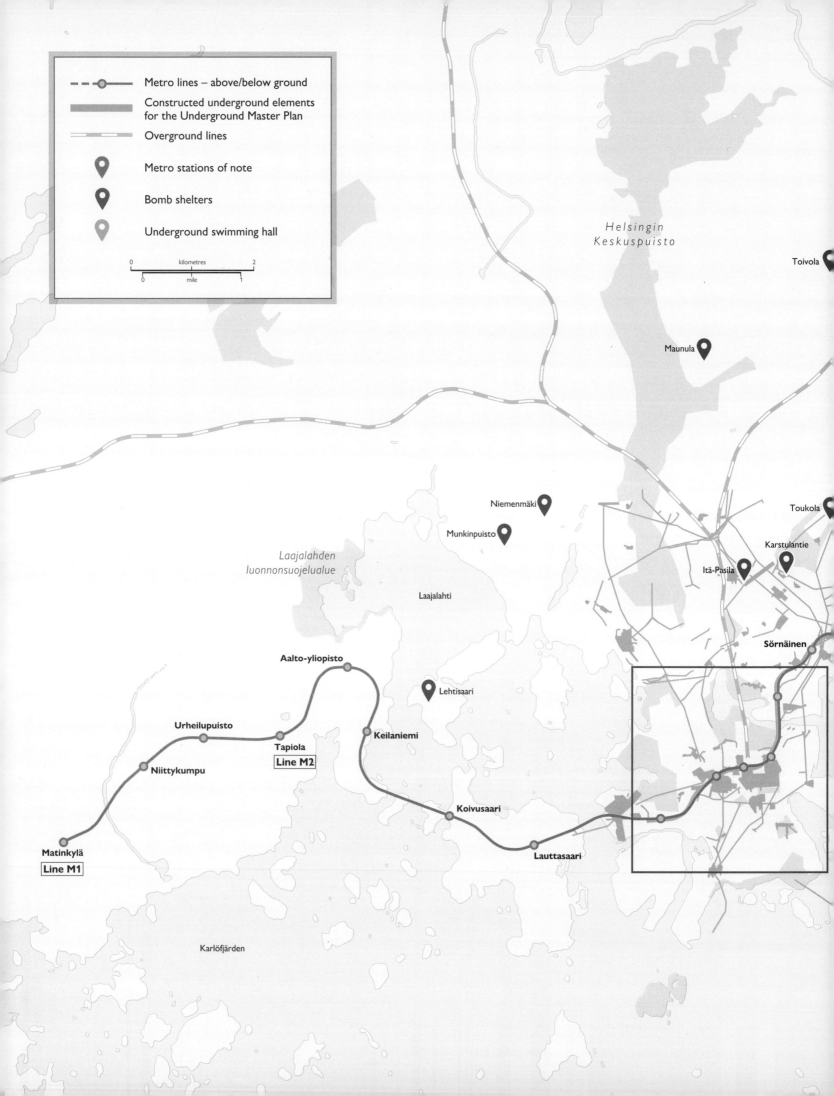

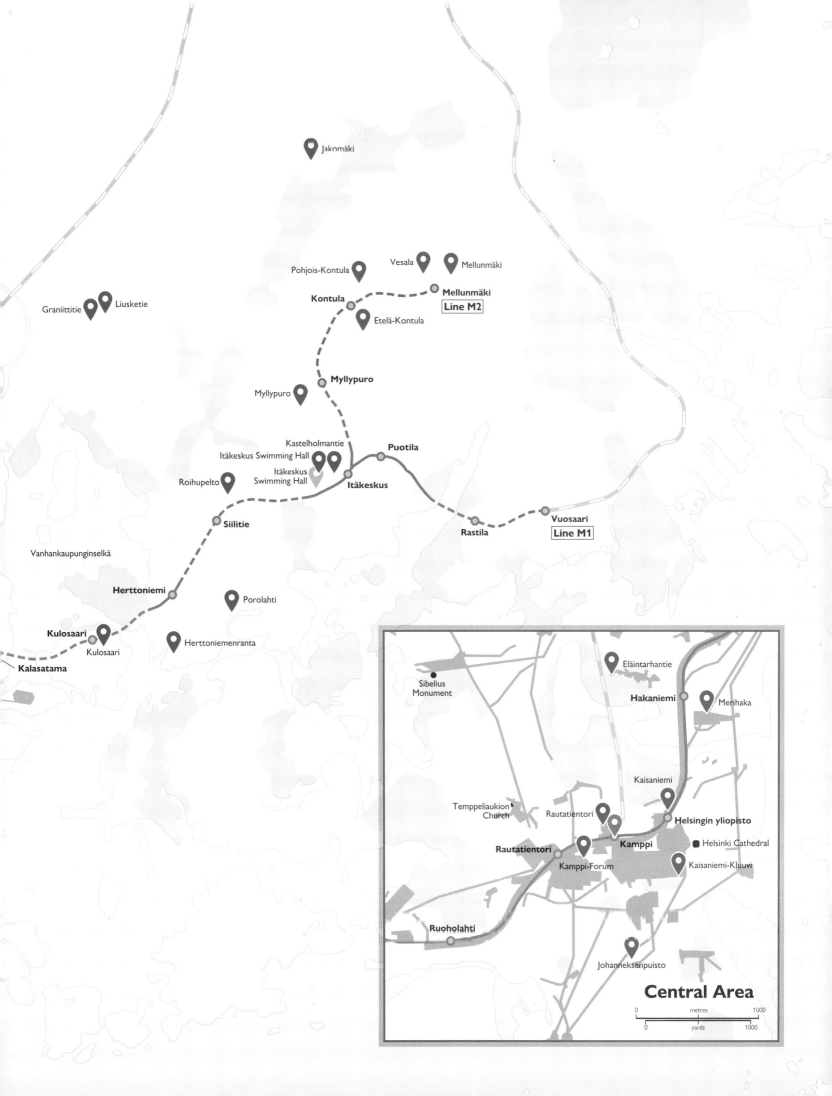

Moscow

Secret subterrania

Earth's northernmost megacity, Russia's capital Moscow, has the largest urban agglomeration on the European mainland, with 13.2 million inhabitants and around an extra seven million in the periphery. It has an instantly recognizable skyline of onion-domed churches, the tallest freestanding tower in Europe and an equally beguiling underbelly.

Traces of Neolithic peoples and Iron Age settlements have been found along the banks of the Moskva River. The site of modern-day Moscow was on many key trading routes and the East Slavic tribes may have settled here in the ninth century. The name of a small town called Moscow can be traced to 1147, but it was razed by the Mongols a few years later.

By the fourteenth century Moscow had been re-established and was becoming quite prosperous. By 1480 it led support for the battle against Mongol domination and Tartar control, and was subsequently named the capital of the empires of Russia and Siberia.

The Kremlin was also being built around this time, with the walls currently standing having been completed in around 1495. By this period Moscow was one of the world's largest cities with about 100,000 inhabitants. However, after its capture by the Tartars in 1571, only 30,000 were still left alive from a population of 200,000. The capital moved to St Petersburg from 1713.

Moscow was restored as Russia's capital again in 1918 following the revolutionary movements of the previous year, and despite the ensuing civil war, the country developed into the first Marxist-Leninist state, the Union of Soviet Socialist Republics, which lasted until 1991. The majority of secret bunkers and tunnels under the city date from the Cold War era, which began at the close of the Second World War.

Moscow's Kremlin

The Kremlin, the fortress at the heart of Moscow, effectively a citadel enclosing churches and palaces, is rumoured to be riddled with tunnels old and new. It also houses what was for many years the tallest building in the city, the Ivan the Great Bell Tower of 1508, which was lifted to its present height in 1600. The meaning of the word *kremlin* as a walled city has been traced back to the 1320s. From 1359, Prince Dmitry Donskoy ruled Moscow for three decades when it is said that many of the underground passages beneath the Kremlin were excavated as routes to the outside world. In later years they were expanded by Russian Orthodox patriarchs as escape routes from the walled fortress in times of siege or attack.

Sixteenth-century ruler Ivan the Terrible hid an entire arsenal in the escape tunnels below the Kremlin, traces of which were unearthed during a 1970s expansion of the metro system. During the rule of Peter the Great (from 1682) there were rumours that Ivan had also buried his library here; said to include unique gold-covered volumes, no trace of such priceless items have been discovered.

A cluster of nine churches known collectively as Trinity Church were developed into Saint Basil's Cathedral during the mid-sixteenth century.

Considered unique at the time, the onion-dome shape of the turrets was said to echo both the nearby bell tower and emulate the flames of a fire. The beautiful edifice is, however, suffering from a degree of instability, as the foundations built in the 1500s have been affected by so many neighbouring subterranean works that the belfry is reported to be leaning noticeably. The most damage to the magnifcient ancient monument of Saint Basil's has resulted from the creation of a huge basement built to store Stalin's biggest military tanks. It was sited at the eastern edge of Red Square in an area once called the Lower Trading Rows, which was demolished to make way for the underground tank park in 1936.

A lined moat was added to the Kremlin, along with more defences, in the following centuries, and the city limits were defined by what is now known as the 16-km (10-mile) Garden Ring road. More attacks, fires, famines, plagues and uprisings decimated the population over the following centuries – not least Napoleon's invasion of 1812, during which anything flammable was burned to the ground. The city recovered sufficiently during the period of the Russian Empire (1721–1917) to house almost two million people by the time of the First World War.

With the exception of the major Moskva River, most of the smaller tributaries in Moscow have over time been canalized, diverted or covered. For example, the Neglinnaya River, which flows from a point in the north of the city near to Savyolovsky station, has been running beneath Red Square and other monuments in the city centre since the nineteenth century. It is channelled inside concrete culverts until its outfall into the Moskva River. Passing under some of the city's most iconic buildings, including the Kremlin, is a new 4-km (2.5-mile) storm drain built from 1974–1979.

The Moscow Metro

Although there was an idea to build an underground railway in Moscow during the days of the Russian Empire, the First World War and subsequent revolution and civil war put paid to activity until 1923. A specific office was then established under the Moscow Board of Urban Railways to study the possibilities of building it.

By 1928 plans were drawn up for the first line, an 11 km (6.8-mile) route from Sokolniki to the centre, which was agreed by the Communist Party Central Committee in 1931.

A full network of ten lines was later included in the Soviet government plan and experts were

ABOVE Once flowing from north to south across what is now central Moscow, the 7.5km (5 miles) Neglinnaya river was given its first culvert as early as 1792. It now runs underground for its entire length, emptying into the Moskva River.

PAGES 184–85 Moscow has arguably the world's most beautiful metro stations: certainly it has more lavishly decorated ones. This tradition was begun when the system first opened in 1935 and continues to those constructed in recent years. Top left: Komsomolskaya (Line 5, opened 1952); Bottom left: Park Kultury (Line 5, opened 1950); Top right: Troparyovo (Line 1, opened 2014); Bottom right: Govorovo (Line 8A, opened 2018).

Metres

- 0
- -10
  - Troparyovo Station (Line 1)
  - Govorovo Station (Line 8A)
- -20
- -30
- Komsomolskaya Station (Line 5)
- -40 Park Kultury Station (Line 5)
- -50
- -60
- Bunker 42 / Rzhevskaya Station (Line 11)
- -70
- Metro-2 (supposedly)
- -80 Park Pobedy Station (Lines 3, 8A; deepest Metro station)
- -90
- -100
- 200 Ramenki bunker (supposedly)

Rzhevskaya Station and Bunker 42

Not only are Moscow's excavations lavish but they are also deep and in some cases secret. There is still controversy over the existence or not of a complete deeper, shadow Metro system but many Cold War constructions which can be confirmed have not only come to light but positively trumpeted. The new private owners of sites like vast Bunker 42 now actively encourage tourists.

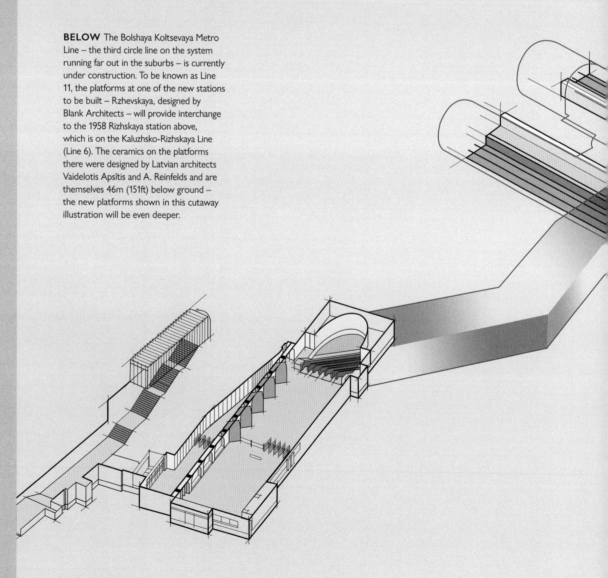

BELOW The Bolshaya Koltsevaya Metro Line – the third circle line on the system running far out in the suburbs – is currently under construction. To be known as Line 11, the platforms at one of the new stations to be built – Rzhevskaya, designed by Blank Architects – will provide interchange to the 1958 Rizhskaya station above, which is on the Kaluzhsko-Rizhskaya Line (Line 6). The ceramics on the platforms there were designed by Latvian architects Vaidelotis Apsītis and A. Reinfelds and are themselves 46m (151ft) below ground – the new platforms shown in this cutaway illustration will be even deeper.

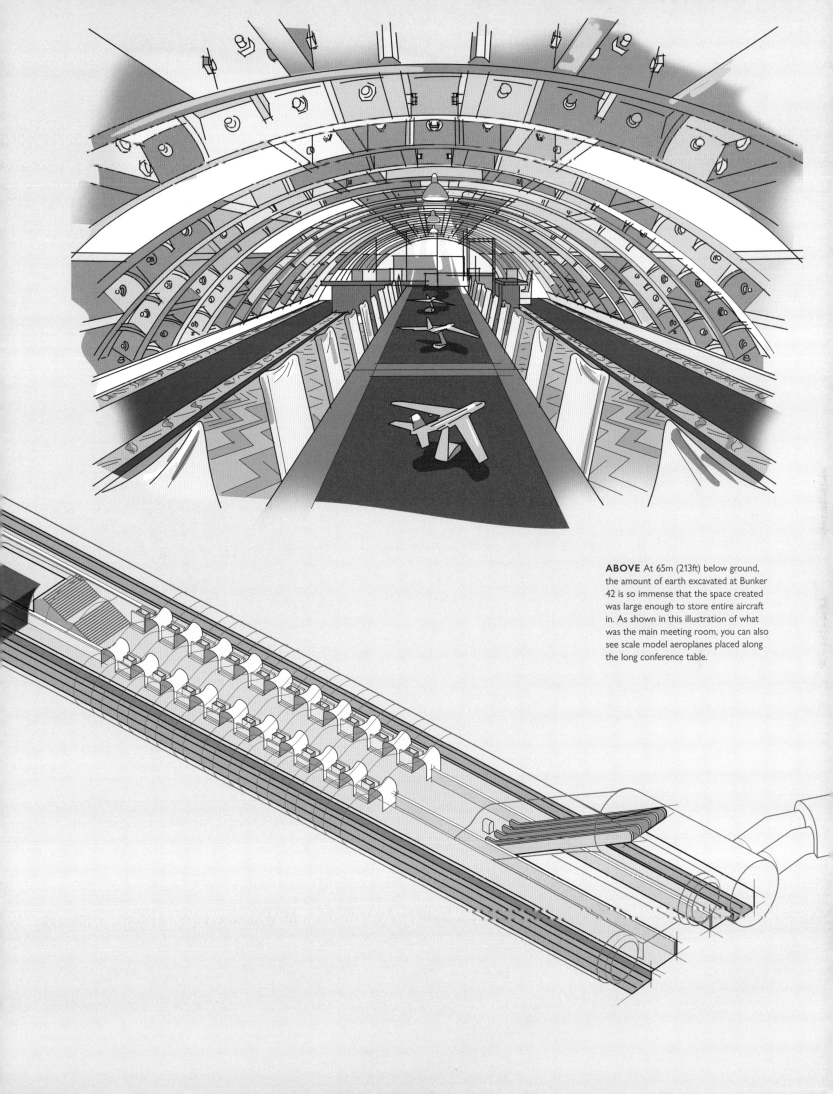

ABOVE At 65m (213ft) below ground, the amount of earth excavated at Bunker 42 is so immense that the space created was large enough to store entire aircraft in. As shown in this illustration of what was the main meeting room, you can also see scale model aeroplanes placed along the long conference table.

called in from Britain's London Transport to help design it. The Brits argued in favour of deep, bored tunnels reached by escalators, as opposed to the more Parisian-style, shallower, cut-and-cover trenches and stairs (or lifts).

The Russians took this advice and almost the entire system was built in deep tunnel. This proved a big advantage when it was required for a secondary purpose – as a shelter against attack. The first line opened in 1935, with the second following in 1937. In later years, the Moscow Metro was specifically engineered to protect those sheltering against the worst effects of nuclear weapons – both blast and fallout. Not only are the tunnels deep, but huge 'blast doors' allow the system to be sealed so that the network can serve as an enormous, interlinked air-raid shelter.

With 437km (272 miles) of track on sixteen lines serving 261 stations, and some of the most frequent headways (distances between trains), the system now supports more than seven million passenger journeys per day. A large proportion of the earliest stations were built in a stunningly opulent style, with marble columns, chandeliers and uplighters, making them among the most grandiose underground spaces on the planet. A number of the most recent additions are also lavishly decorated, albeit in modern styles. They include Troparyovo on Line 1 (2014) and Govorovo on Line 8A (2018). At a depth of 84m (276ft), Park Pobedy, (1986), is the fourth-deepest metro station in the world, and has Europe's longest escalators.

Alongside the onset of frosty relations with the West, and the understanding of the benefits that tunnels under a city can bring, the Moscow administration is reported to have agreed to the construction of what is now known as Metro-2. At 50–200m (165–655ft) below the surface, at least 25km (15 miles) of secret tunnels, and possibly much more, were allegedly ordered by Josef Stalin to link a deep, underground command post beneath the Kremlin, with at least three others outside the city and another near to Moscow State University. The existence of these tunnels has not been confirmed by the Russian government but, according to leaked US Intelligence, there are three routes out of the city and local 'diggers' (urban explorers) claim to have photographic evidence of them.

At least one of three lines, built in the 1970s, is thought to have had a direct rail connection to Line 1 of the public system. One Metro-2 'station' is said to provide access to Ramenki underground bunker (see below). Another route, supposedly built during a second phase in the 1990s, is claimed to run 28km (17 miles) out from the city centre to the former military airport at Vnukovo.

Meanwhile, the official public Metro has been undergoing another major expansion phase since 2012. In the decade to 2022

LEFT Opened as a strategic nuclear facility in 1955, Bunker 42 contains around 7000m² (75,000ft²) of space and as well as an aircraft hangar, there were also communication and command centres, accommodation, offices and restaurants. Now a museum of the Cold War, this restaurant gives some idea of the vast spaces created.

RIGHT There are many corridors like this inside Bunker 42, some of which led to the nearby Taganskaya Metro station, not just for pedestrians to enter: trains also ran into it from the Metro tracks to aid construction of the site.

almost 150km (90 miles) of new tunnels will have been constructed using twenty-four different tunnel boring machines. Several lines are also being extended at the time of writing and the brand new Nekrasovskaya Line, which already has the first section open, should be fully ready by 2020. Referred to as either the Nekrasovskaya Line or Line 15, the first 14.4km (9 mile) section was opened in 2020 and will eventually reach 22.3km (14 miles). Construction was quite unusual for Moscow: it was built as a 10m (33ft) diameter double-track tunnel – rather than the single bore tunnels mostly used elsewhere on the Moscow Metro.

Moscow's bunkers

There are now said to be so many underground spaces beneath the city that the entire population of Moscow could be sheltered in the event of a nuclear attack. Given its size – in excess of thirteen million people – that requires a lot of space! The bunkers can be categorized in four major groups: basement and Metro bunkers – the main places of shelter for civilians – and sphere and Metro-2 bunkers, both of which were used by the military. The basements are the oldest and, as the name suggests, were essentially shallow excavations built relatively cheaply and quickly beneath existing buildings and parks. They were scooped out and covered over with concrete. A sphere-type bunker is characterized by a spherical concrete shield that acts as a buffer against a nuclear blast. Spheres vary in size and age – in the newer ones, the concrete shield is part of the original construction and includes shock absorbers to protect those inside. Earlier sphere bunkers – generally less deep and built during the Cold War period – had their spherical conrete protection added afterwards, on top of existing structures. One example of a Metro-style shelter is Tagansky Bunker 42 complex, which now functions as a museum.

In 1992 US *Time* magazine reported the existence of a bunker so large it was effectively the size of an underground city. According to the magazine, its source confirmed that construction began in the 1960s. The site, near Ramenki in the southwest of Moscow and close to Moscow State Univeristy campus, was completed by 1970 and was to be used by a science association. The structure is theoretically up to 200m (656ft) deep, providing accommodation for 15,000 people, which would make it the deepest and biggest bunker in the city, if not the world. It is supposedly connected to the Metro-2 system and various websites claim an entrance to 'Ramenki-43', although that address is currently occupied by two supposed emergency agencies, the Militarized Rescue Squad 21 and the 1st Paramilitary Rescue Squad.

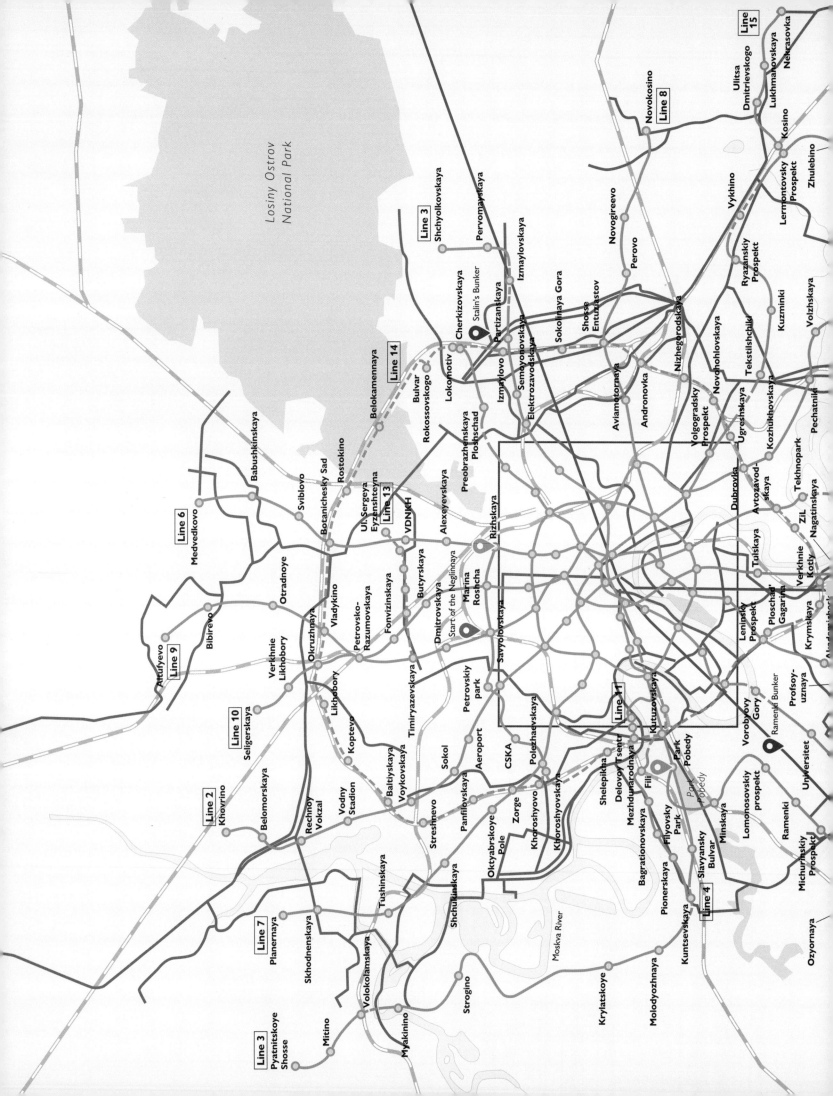

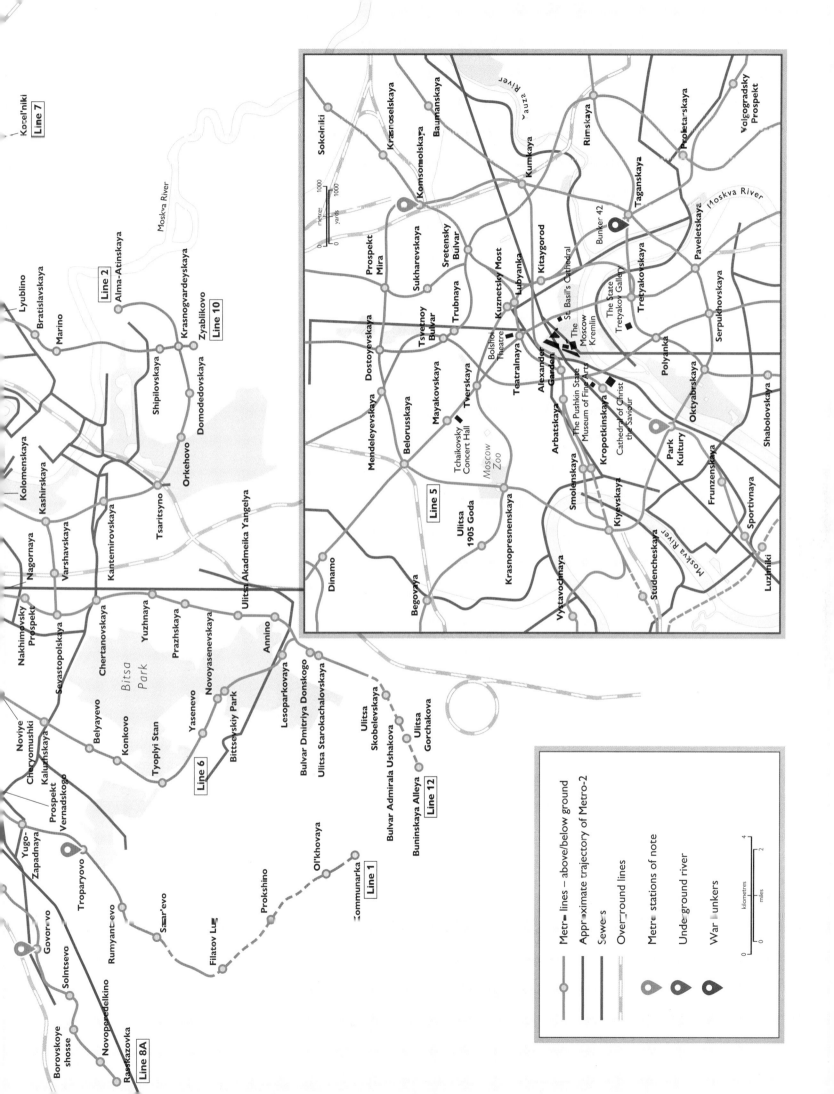

Asia and Oceania

A 1923 photograph documenting the construction of a tunnel in Hyde Park as part of Sydney's City Circle underground railway.

Mumbai

City of seven islands

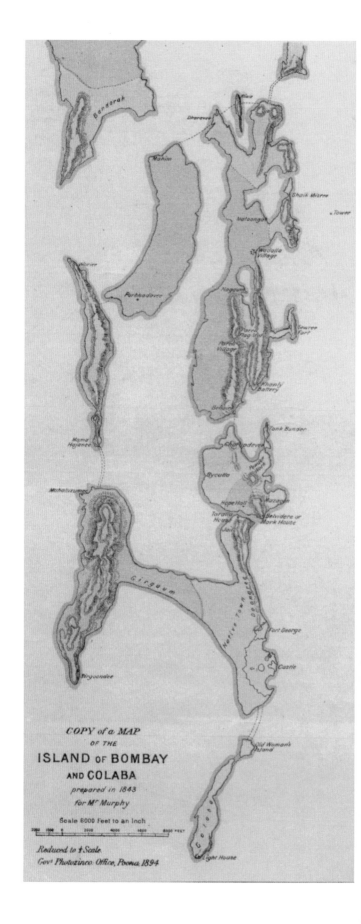

The city once known as Bombay, and now the capital of the Indian state of Maharashtra, was built on an archipelago of seven islands on the country's west coast. It is home to twelve million people with more than twenty-one million living in the broader metropolitan area.

There is evidence that the oldest parts of Mumbai were occupied in prehistoric times by the Koli people from Gujarat. During the sixteenth century, the Islands of Bombay were part of the Portuguese Empire and, in 1661, were rented to the British East India Company that formed in 1600 for commercial trade around the Indian Ocean. The firm eventually took over land in what later became India and played a pivotal role in colonizing Southeast Asia.

Reclamation

Several ambitious land-reclamation projects were undertaken from 1782, to join the seven islands into one via various landfills and causeways. Among these, the Hornby Vellard project was a long roadway on a raised embankment between the biggest land area, the 'H-shaped' Isle of Bombay and tiny Worli Island just 0.3km (0.2 miles) to the north. The road measured 4.2km (2.6 miles) north–south, with a 4.5-km (2.8-mile) east–west section. Completed in 1784, the causeway shielded the original central lagoon of Bombay from regular flooding. The 283 hectares (700 acres) of swampy area between the reclaimed areas and islands was progressively drained, further causeways were built and, by 1838, the former Isle of Bombay was no more. Once joined to the mainland, and to the other islands, its population expanded rapidly.

Ancient tunnels

A 240-year-old tunnel running for 1.5km (1 mile) from St. Georges Fort to beneath St George's Hospital was recently rediscovered. It was used for moving weapons and injured troops, and archaeologists suspect it may also be part of a network that once linked via other tunnels to Churchgate, Blue Gate and the Apollo Bunder (now known as Wellington Pier with its landmark arch monument, the Gateway of India).

Yet more ancient tunnels date from the mid-nineteenth century – for example, a secret bunker recently dicovered beneath Raj Bhavan, the former

ABOVE Artist's impression of part of the Raj Bhavan bunker which in total covers 1,400m² (15,000ft²).

LEFT The original seven islands of Bombay Harbour, mapped here in 1894, were in just a few years all joined to the mainland (in the top left of the image) by a network of causeways and land reclamation.

home of the governor of Maharashtra, which was created as an army barracks and covers around 450m² (4,850ft²). The precise date of construction is unknown, although it is believed to have been later than 1885.

Transportation in Mumbai

While the city was still known as Bombay and was part of the British Empire, a large network of railways developed across the Indian subcontinent. The city was so well served that the population grew rapidly along what are now the commuter lines of Mumbai Suburban Railway (Western, Central and Harbour lines), which provide more than seven million passenger journeys a day. A 1.3-km (0.8-mile) tunnel between Parsik and Thane, which opened in 1916, is one of the oldest and longest in India.

A tram network helped growth too, but it was closed in 1980, at precisely the time the city was expanding rapidly. Today Mumbai faces chronic traffic congestion on a daily basis, so the decision was finally taken to invest in a modern mass transit network. Very much in its infancy, the Mumbai Metro has just one line so far, which is 11.4km (7 miles) long. Opened in 2014, and mostly elevated, this is the start of what will be a much bigger network of eight lines, currently under construction. When finished, 24 per cent of the network will be underground. This will mostly comprise all of Line 3, which will run 33.5km (21 miles) and serve twenty-seven stations between the southern tip of the city at Cuffe Parade in the Central Business District and Santacruz Electronic Export Processing Zone (SEEPZ) in the north. There are plans to extend the network further, to 235km (146 miles) by 2025, although these are currently behind schedule owing to the difficulties of building in a crowded and ancient city. When built, however, this next phase could almost double the network to fourteen lines in total.

Beijing

Created by hand

Home to twenty-one million people with three million more on the periphery, the Chinese capital has rocketed up the ranks to become one of the most populous urban spaces on Earth.

Although archaeological evidence suggests that human ancestors lived in this area hundreds of thousands of years ago, the Beijing of today has roots going back around three thousand years, to the founding of Ji City under the Western Zhou dynasty. In the eight hundred years leading up to the twentieth century, the city functioned as a 'capital' at least half a dozen different times during the rule of thirty-four different emperors.

During the Ming dynasty (1368–1644) a series of walls with fortified gates was erected around the city's perimeter. The remnants of these, including the 20m-thick (66-ft) Inner City Wall, which was 24km (15 miles) long and 15m (49ft) high, have endured to provide the shape of modern central Beijing. The Forbidden City palace complex, containing hundreds of buildings, was constructed in the early fifteenth century and is now a World Heritage site.

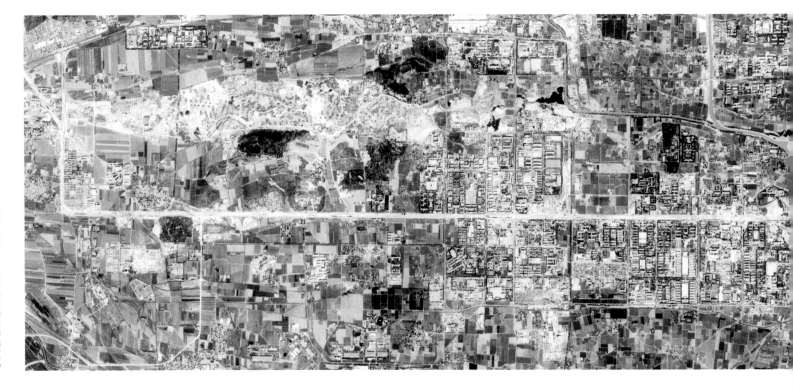

Beijing's subterranean city

Despite the incredible advances in civilization that China has demonstrated in its long history, it is arguably the rapid changes since the civil war of 1945–1949 and, more specifically, the introduction of economic reforms in 1978, that have most radically transformed it.

In the decade leading up to 1978, the country feared the threat of war with its neighbour, the former Soviet Union. In 1969, Chairman Mao Zedong ordered his people to prepare for possible nuclear attack by digging their own tunnels, an instruction that Chinese citizens embraced. Up to 300,000 of them in Beijing alone excavated an unprecedented number of up to 10,000 bunkers beneath the city, mainly using hand-held tools.

For ten years, during some of the most intense periods of the Cold War, occasionally assisted by the army, but mainly by their own strength of will, the Beijing people squirrelled away at carving and sealing and interlinking up to 85km² (33sq mi) of an underground city. Known locally as the Dìxià Chéng, the network was three levels deep and had as many as nine hundred separate entrances. The tunnels included links to every major government office and would have been capable of sheltering the majority of Beijing's six million people – the number of inhabitants when work began in the 1960s. Contained within the depths were restaurants, factories and storage facilities, as well as theatres, sports halls and even a farm for basic crops such as mushrooms.

With the exception of mines and underground railways, it is surely the most impressive of all human-made subterranean creations. As international relations thawed in the 1980s, property prices above ground began to soar and a number of people who couldn't afford expensive apartments began to occupy the spaces below ground. The inhabitants came to be known as the Rat Tribe. Recent attempts have been made by authorities to clear them, but most are still there.

A few entrepreneurial citizens, and even large property developers, have repurposed remnants of the Dìxià Chéng. Some of the underground spaces have even been converted to the basements of shopping malls (along Wangfujing Street). Other spaces are used by tourists: Chang An Grand Hotel now goes six storeys down utilizing part of the Dìxià Chéng works.

There is speculation that the tunnels run much further out from the city than would benefit simply those sheltering from localized attacks. Dissidents have claimed that much wider, tarmacked tunnels link far-flung army barracks and tank storage facilities and that they run as far as

BELOW 1967 US Government spy satellite photo of the construction works on the route of the first line of the Subway. The demolition of the historic wall can be clearly seen during the invasive cut'n'cover technique. Nowadays almost all new tunnels in Beijing are bored.

neighbouring cities. There are also suggestions that hidden underground military road and rail networks exist right across the country.

Ambitious subway plans

The first line of Beijing's underground railway opened in 1969. There had been discussion of a metro system since the early 1950s. Studies had shown that the Moscow Metro had the ability to move the populace while also doubling as a military asset. But a 1957 plan for six lines totalling 172km (107 miles) and with 114 stations – made with Soviet guidance – never came to fruition.

Construction finally kicked off in 1965, on a 21-km (13-mile) east–west line that caused much controversy as parts of the city's ancient inner walls and gates had to be demolished. Although the line was technically finished in 1969, owing to engineering issues only an initial 10-km (6-mile) stretch was able to open in early 1971, being extended piecemeal in the next year. By 1981 its length had reached 27.6km (17.1 miles) serving nineteen stations, and control passed to the Beijing Subway Company, which had plans for many extensions. In just three years it had completed a second route, which ran in a horseshoe shape beneath the line of the former Ming dynasty city wall.

In 2001, the announcement that Beijing had won the right to stage the 2008 Summer Olympic Games kick-started the most rapid expansion of any metro system on the planet. Initially a bold declaration of building 'three ring, four horizontal, five vertical and seven radial lines' shook the construction world, but in a joint venture with Hong Kong's transport provider Mass Transit Railway, the plan for all nineteen lines covering 561km (349 miles) has essentially been achieved.

In 2008, yet more ambitious plans announced a further extension of the Beijing Subway by 2022, running to 1,000km (620 miles) of track in total – 62 per cent of it in tunnel – and with twenty-two lines serving an as yet undisclosed number of stations, though it is likely to amount to more than six hundred.

ABOVE For the 2008 Olympic Games Beijing built a vast venue park, including the 91,000 capacity 'Bird's Nest' stadium. A huge, man-made green area called Olympic Forest Park was opened simultaneously and to get visitors to the sites a new Subway was constructed. Some stations on the new Line 8 included flagship features like South Gate of Forest Park (pictured) with an impressive platform roof designed by Foshan Xinjing Industrial Company.

RIGHT The new Subway stations were built so fast that many of them are lacking in the style and panache of others around the world. Here, a very utilitarian looking Huoqiying station on Line 10, opened in 2012.

Rail expansion is not limited to metro-style operations. Built for the easier passage between the two mainline stations Beijing and Beijing West a 9-km (5.6-mile) cross-city rail tunnel opened in 2015. Since 2017 it has also been utilized by the new Sub-Central Suburban line.

Sanitation and water needs

The Chinese authorities have tackled the city's sewage requirements on an equally monumental scale as its transport, though critics argue this has been less successful than the Subway. Even as late as the 1990s Beijing's sewage system was woefully inadequate. New drains and pipes totalling 3,000km (1,900 miles) have now been built and, since 2000, a new sewage-processing plant has opened every year, but there is still much progress to be made.

Added to this, Beijing is short of water – so much so that there has even been talk of moving the entire city to an area with higher levels of rainfall. Depsite there being billions of litres of fresh water available, the volume falls some 16 per cent short of quenching the needs of this gigantic city. Deeply buried water is being sucked out of the ground so quickly that it's causing land levels to sink. Such is the shortfall, that desalinated water may be pumped in underground from the coast, over 150km (93 miles) away.

Ancient treasures

Although there is little evidence of ancient tunnelling beneath the Forbidden City, two underground storage facilities (built in the 1980s and 1990s) are being expanded to 29,000m^2 (310,000ft^2) to help the site's museum to keep more than one million precious artefacts in climate-controlled conditions. At the moment artefacts are scattered in various buildings throughout the city, but this new subterranean space will allow one of the storage facilities to function as a new furniture display. A tunnel connecting this expanded warehouse space with the repair workshop will avoid any need to convey delicate items in the open air from one part of the site to another.

Tokyo

Meeting under a megalopolis

With more than thirty-eight million inhabitants, and covering 13,500km² (5,200sq mi) of the Kanto River floodplain, Greater Tokyo is the world's most populous metropolitan area. With a GDP of US$1.8 trillion dollars, it is also the world's most productive city. It is located on Tokyo Bay, Honshu, the largest of the four main islands that make up Japan. Destroyed and rebuilt many times as a result of earthquakes, the city's historic heart was originally known of as Edo, which translates roughly as 'estuary', although it was not yet the capital of these great islands. Modern To-Kyo (Eastern City) was founded here in 1868 by the Japanese Emperor Meiji, who transferred his court to here from Kyoto.

Forces of nature

The geographical location of this gargantuan city makes it prone to the effects of many natural phenomena that have influenced construction both above and below ground. It is particularly susceptible to earthquakes, tsunamis and flooding.

Due to the number of tremors and an insatiable appetite for high-rise buildings, the foundations of many Tokyo skyscrapers are exceptionally deep in order to anchor them to the bedrock. New buildings have to navigate the already complex mesh of utilities, sewers and subways to pierce far enough down to make the building above ground safe. Tokyo now has more than three hundred buildings over 100m (980ft) in height, and in the most recent and violent quakes, not a single building of any height collapsed.

Urbanization entirely surrounds Tokyo Bay. The driving distance between Kisarazu on the eastern side and Kawasaki on the western shore used to take around ninety minutes via downtown Tokyo. In the 1970s a plan was hatched to build a combined bridge and tunnel to connect the two sides of the large bay. In 1997, after almost a decade's construction and US$11 billion, the Tokyo Bay Aqua-Line opened, cutting the 100-km (62-mile) drive to just 14km (8.7km) over and under the sea. The tunnel is up to 45m (148ft) deep and at 9.6km (6 miles) it is the longest undersea car tunnel in the world. The Yamate road tunnel opened in the 2000s is even longer at 18km (11 miles). Planned as an elevated expressway in the 1970s, it was thwarted by environmentalists but returned in tunnel form twenty years later. It took fifteen years for the first section to be ready and another decade for it be fully opened.

ABOVE Artist's impression of the enormous concrete pillars and roof of the G-Cans flood defence chambers.

Water presents another problem for the metropolis: it rains a lot. The summer monsoon can deposit up to 10cm (4in) rainfall per hour. A major flood during 1991 sunk 30,000 homes. In response, a radical plan was hatched to help drain off excess water in severe storms. The Metropolitan Area Outer Underground Discharge Channel – sometimes called 'G-Cans' and dubbed the Underground Temple – consists of five gargantuan subterranean concrete silos 50m (164ft) below Kasukabe, 45km (28 miles) north of Tokyo. Linked by 6.5km (4 miles) of tunnel, each is capable of holding about 670,000m^2 (7,200,000ft^2) of water. Construction on the project started in 1993 and has taken over thirteen years and cost US$2 billion... but it works! At least half a dozen times a year, the vast tanks begin to fill with storm water. The system stores and then pumps and drains the water safely into the Edo River which flows out into Tokyo Bay.

Modern attempts to deal with unwanted water were preceded by sewage and plumbing around the more ancient settlements such as Edo and Kodemmacho. Some claim Edo was the world's largest city by 1700 (the 'commoner' population was counted in 1678 as 570,361, and by 1721 could have topped one million people), so something had to happen with the unwanted waste! Scholars have cited the removal of waste by hand and ship, but recent archaeological works have unearthed the 1864 Kanda Sewer – some 65km (40 miles) of square wooden pipes and channels. Later, water from the Tama River was also pumped down to the city from nearby hills 43km (27 miles) away.

Thanks to Japan's mountain ranges and limited flat land for urban space, real-estate prices in cities are at a premium. Surface car parks are almost unheard of now and most

Metres	
0	Edo (ancient Japanese capital)
-10	
	Bike vaults
-20	
	G-Cans flood control system
	Subway tunnels (average)
-30	
	Higashi-Shinjuku Station (Ōedo, Futukoshin Lines)
-40	Kanda sewer
	Roppongi Station (Hibiya, Ōedo Lines; deepest Toei Subway station)
	Aqua-Line
-50	MAOUDC water tunnel
-60	
	Gas pipes
-70	
-80	
-90	
-100	

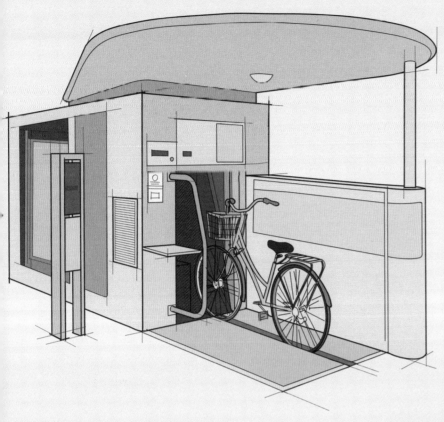

Bike Vaults

Tackling the issue of real estate at surface level, even Tokyo's thousands of bikes are forced underground. About 15 per cent of the daily commute is by bicycle which in a city this size would represent many thousands of bikes cluttering the footpaths. The solution was of course to store them in these giant sub-surface tanks. Known of as EcoCycle these illustrations show how a typical structure can bury the bikes from very small street level kiosks.

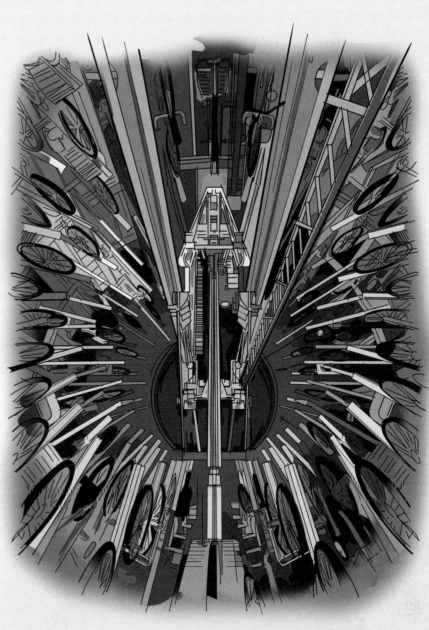

modern buildings have several basement levels for vehicles. But even securing the cities many bicycles was taking up valuable space, so a recent solution has been the bike vault. Cylindrical shafts are sunk to a depth of up to 13m (43ft) and specialized equipment installed that can automatically pull the bike in from a street access point, and robotically stash it deep in the vault. When a cyclist returns, their bike is rapidly retrieved and raised back to the surface for immediate use.

Urban rail network

Being a vast metropolitan area, and a country that loves railways, means that the Greater Tokyo area has one of the densest concentrations of urban rail transit on the planet, and much of it runs below ground. Besides an enormous mesh of suburban commuter lines, the metro system itself is provided by two separate organizations: the privatized Tokyo Metro (180km/112 miles of routes serving 179 stations on nine lines) and the state-run Toei Subway, which serves 108km (67 miles) of routes and has ninety-nine stations on four lines.

What is now called the Ginza Line (operated by Tokyo Metro) was the first to open, in 1927. It claims to be the oldest subway in Asia, although there was a post-office railway under Tokyo's main station that started in 1915, but that was for mail and not passengers.

The initial success was a portent for the popularity of the idea. It was said that passengers were prepared to wait for up to two hours to board a train for journey that would last just five minutes. Another line, operated by the Tokyo Rapid Transit Company, opened in 1938. Two operators merged after the Second World War under the name Teito Rapid Transit Authority, which was controlled by the city.

More lines opened in the following decades, culminating in 2008 with the Fukutoshin Line, which is the deepest one in Tokyo, reaching 35m (115ft) below the surface at Higashi-Shinjuku station). It features the ability to run express trains that skip certain stops.

Oddly, given its long service, the system has few abandoned stations but Manseibashi on the

ABOVE This aerial view of Greater Tokyo shows only a slither of the world's largest urban expanse. The amount of services and transportation buried beneath this conurbation is simply mind-boggling. Generating two trillion dollars of GDP annually, Tokyo covers (depending on where the line is drawn) over 13,500km^2 (5,200sq mi).

RIGHT The Tokyo Bay Aqua-Line, which opened in 1997, is a combination bridge-underwater tunnel. The photo shows the area where the elevated causeway dips below the water level into the long undersea tunnel. The artificial island, which also houses the tunnel ventilation system, is known as 'Umihotaru' (sea-firefly) and includes toilets, lounges, restaurants and amusement arcades.

first line, in the Chiyoda area, opened in 1930 and closed a year later when the line was carried over a nearby river.

Tokyo's hidden city

More recent Tokyo lines, such as the Hibita (Line 6) and the Ōedo (Line 7) seemed to have been constructed much faster than might have been expected, leading some to suggest they were utilizing parts of tunnels that already existed.

Writer Shun Akiba has postulated that there is a secret hidden city buried beneath modern Tokyo. Accidentally spotting what looked like an error on architectural plans, Akiba started investigating what appeared to be an uncharted mystery tunnel beneath the subway. The more he discovered, the more baffling the situation became. He started noticing odd tunnels leading off the existing metro lines that are not shown on any maps or construction records.

Although the official tally of rapid transit lines in Tokyo is around 250km (155 miles), Akiba suggests there may be up to 2,000km (1,240 miles) of undeclared tunnels beneath the city. He cites buildings constructed over huge pre-existing basements for which there are no published plans – for example, the eight lower levels of the National Diet Library, which were not built by the institution itself. Abika says: 'It seems likely that the subterranean complex was prepared for a possible nuclear attack.' However, no official sources will confirm or deny the existence of these mysterious places.

World's deepest foundations

At 634m (2,080ft), the Tokyo Skytree is the tallest structure in Japan, but it's set to be dwarfed. So confident are Japanese architects in modern technology's ability to withstand earthquakes, there are now plans for the world's highest megatall skyscraper. Expected to be built on an archipelago of reclaimed land off Tokyo Bay, the Sky Mile Tower will be 1,700m (5,577ft) tall with 400 floors and home to 55,000 people. Naturally, it will have the world's deepest foundations, although some patience is required for such lofty ambitions: it is not anticipated to be finished until the year 2045.

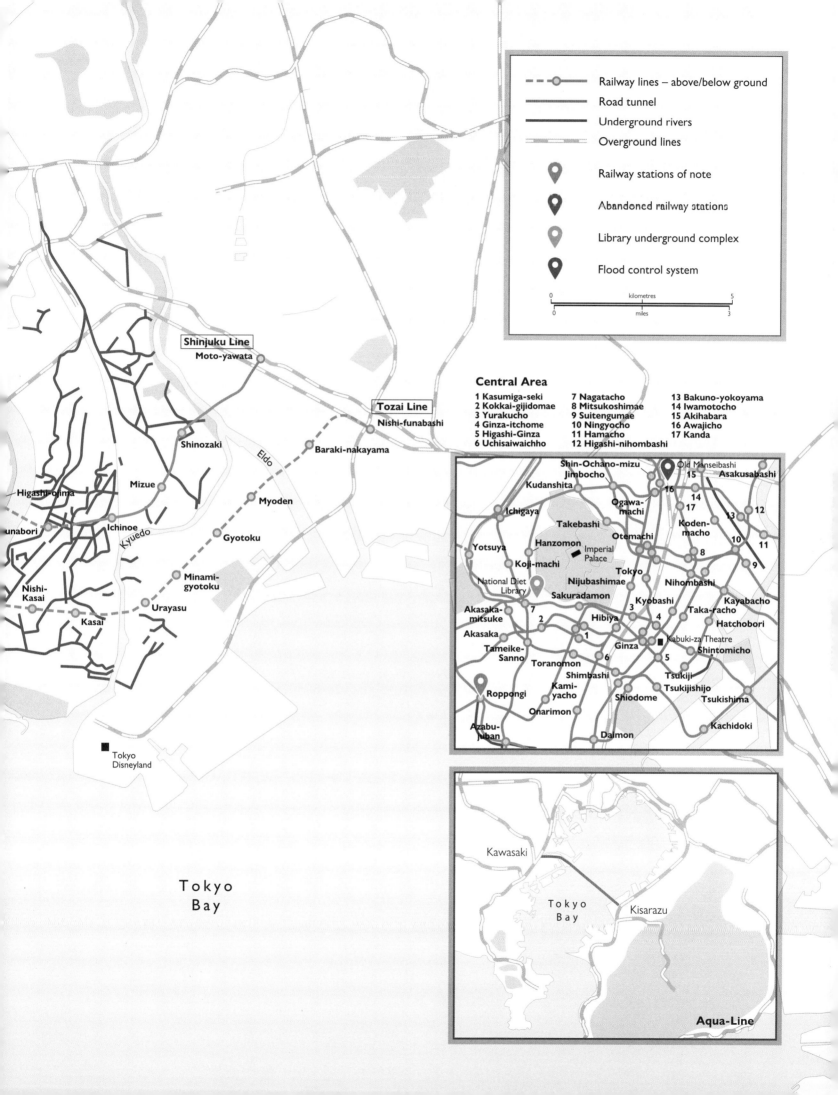

Sydney

Roads to nowhere

Located on a natural harbour on Australia's east coast, Sydney has an urban area of around 4.4 million people, growing to well over five million if counting the sprawling metropolitan suburbs surrounding it. This figure makes the city Australia's largest, and represents more than 20 per cent of the country's population.

The area in which Sydney is now located has been inhabited by the indigenous peoples of Australia for at least 30,000 years – some claim that Aboriginal tools found in the sediment can add up to 20,000 years more to that date. British colonists under Captain James Cook were the first Europeans to map the coast in 1770, having landed at Botany Bay. Eighteen years later a penal colony was founded at Port Jackson.

Early infrastructure

Fresh water was an issue for the early colonists, but a tributary to Sydney Cove played a role in the settlement being based there rather than at Botany Bay. Just three years after the founding of Port Jackson in 1788, the waterway – known as Tank Stream – was being managed and was quickly bridged. The water was so polluted by the 1820s that it was covered at Bridge Street and converted to an official sewer in the 1850s. British civil engineer John Busby replaced Tank Stream progressively between 1827 and 1837, by employing convicts to dig a 3.5-km (2-mile) tunnel – the first in modern Australia's history – to bring fresh water from a pond in what is now Centennial Park. 'Busby's Bore' has heritage status today. A need for proper municipal planning inspired the creation of Sydney Corporation in the same year that Sydney gained city status in 1842.

Although the first railways in Australia had been constructed twenty years earlier, they did not reach the edge of downtown Sydney until 1855, arriving at an area known as Cleveland Fields. The original single-platform stop was named Sydney Terminal and subsequently expanded (eventually to fourteen platforms) and relocated to its current position on Eddy Avenue, a few hundred metres to the north, in 1906. It is now known as Central railway station.

Sydney Harbour Bridge

As the city grew exponentially, plans kept evolving for what became the vast Sydney Harbour Bridge project – a crossing first dreamt of in 1816. The plan was to accommodate rail tracks to link existing Sydney Central station across the bay to tracks on the north shore at Milsons Point. Construction took nine years, partly because the foundations on either side needed to penetrate deeply into the rock in vast underground caverns. The bridge was opened in 1932.

The ability for trains to pass through, rather than terminate at, Sydney Central gave rise to major rearranging and rebuilding that required whole new sections of railway, including key lines from Sydney Central. The tracks were extended first underground (via Town Hall and Wynyard) then to surface and cross the new bridge to join the existing northern branches on the other side at Milsons Point.

LEFT A member of the Royal Australian Engineers exits a well at Victoria Barracks after helping clear and restore Busby's Bore, 1975.

An early manifestation of the Harbour Bridge realignment was the construction of the City Circle. Starting at Sydney Central, this short line dives underground to form a small loop beneath the Central Business District. It serves six stations including Sydney Central: Town Hall, Wynyard, Circular Quay – where it effectively pops up into open air – St. James and Museum. Although it took some time to complete, its tracks link to the city's surface commuter lines, effectively giving Sydney Trains (formerly CityRail) a network stretching over 800km (500 miles) and serving 178 stations.

Proposed by engineer John Bradfield, the loop was always supposed to be the centrepiece of a full underground metro network. Although several tunnels were built, they were never used for their intended purpose. An abandoned, half-built tunnel beneath St James is now filled with water and dubbed the 'lake'. Its inner two platforms, built for underground services to the eastern suburbs, have never been used. This line was to enter the city from Darling Harbour, and then run in tunnel from Town Hall to O'Connell Street and St James. The platforms for this service were built at Town Hall station and are now used by the Eastern Suburbs Railway. Another abandoned tunnel led from Wynyard onto and over the Harbour Bridge. It was originally used by trams between 1932 and 1958. Across the water, there are 500m (1,640ft) of never used tunnel at North Sydney heading to Mosman.

Reaching the suburbs

The Eastern Suburbs Railway itself was built during the 1970s. It had been three decades in planning but when it was completed, in 1979, it only managed a short shuttle between Sydney Central

Sydney Opera House Car Park and Macquarie Lighthouse

Owing in part to its size and in part to its comparatively long history, Sydney has some of the most extensive underground infrastructure in the Southern Hemisphere (topped only by São Paulo, Brazil and Buenos Aires, Argentina). While it has not developed such extensive subway systems that cities of a similar scale are equipped with, it does contain many abandoned tunnels from attempts to build such a network, plus three tunnels which were once used by the former Pyrmont Goods freight line. Of all the subsurface car parks in the city the one beneath the Opera House is the biggest and most impressive.

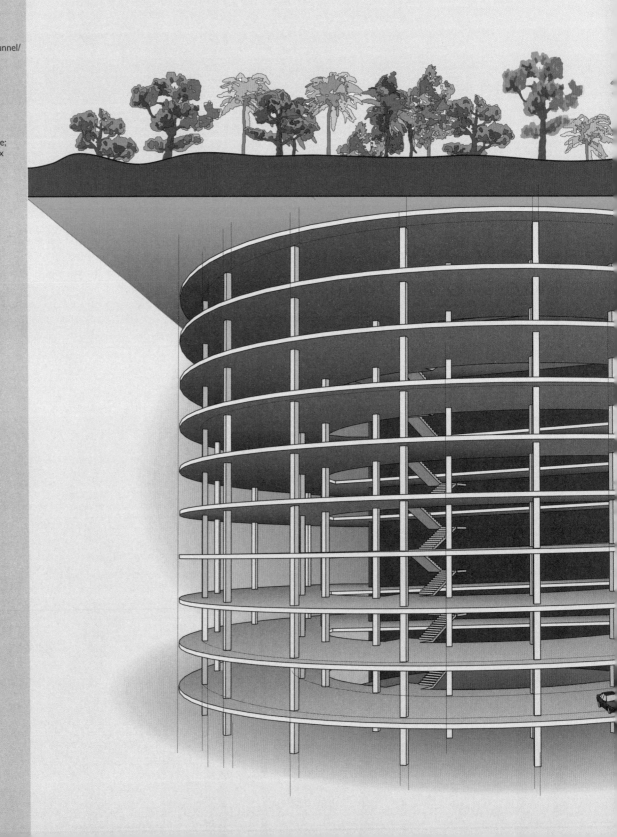

Metres

- 0 — City Circle
- St James 'lake'
- North Fort Plotting Room
- 20 — Cross City Tunnel / Victoria Barracks water shaft
- Sydney Harbour Tunnel/Lane Cove Tunnel/ Busby's Bore
- 30 — Eastern Distributor
- North Ryde Station (Northwest Line; deepest Metro station)/WestConnex
- Sydney Opera House Car Park
- 40 — Metro City & Southwest tunnels
- Metro Northwest tunnels (Epping–Bella Vista) at West Pennant Hills

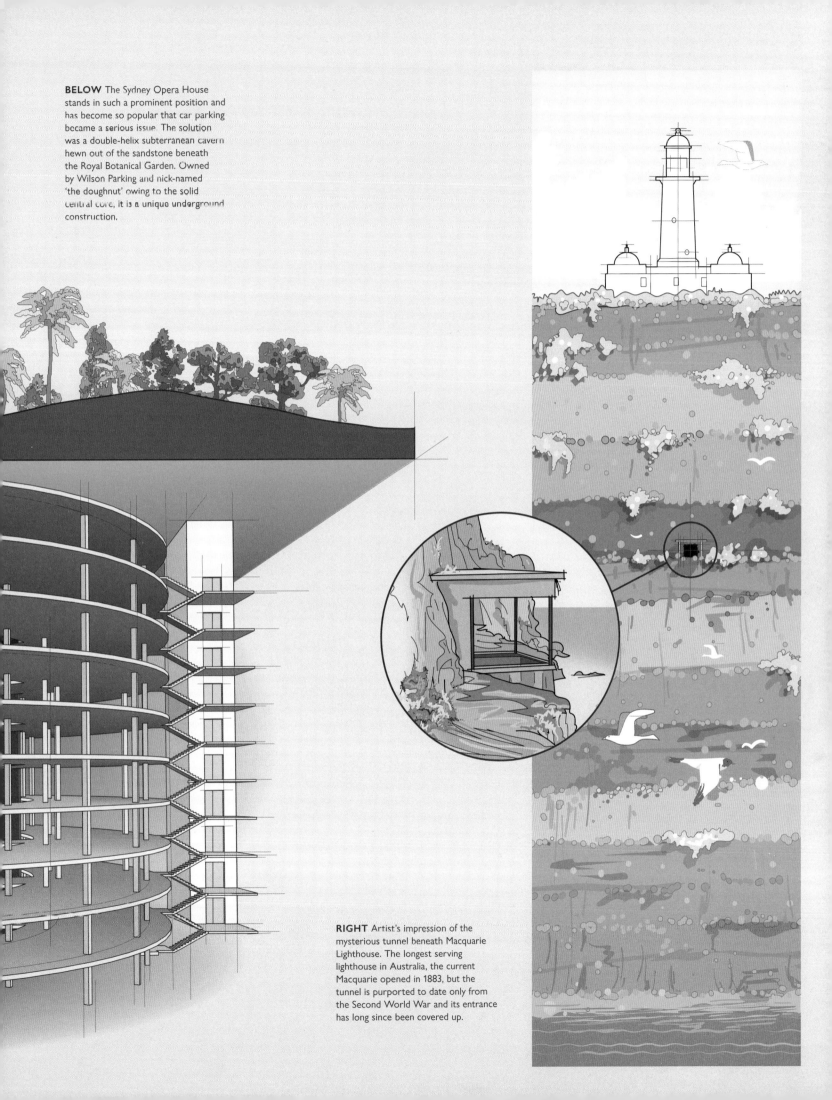

BELOW The Sydney Opera House stands in such a prominent position and has become so popular that car parking became a serious issue. The solution was a double-helix subterranean cavern hewn out of the sandstone beneath the Royal Botanical Garden. Owned by Wilson Parking and nick-named 'the doughnut' owing to the solid central core, it is a unique underground construction.

RIGHT Artist's impression of the mysterious tunnel beneath Macquarie Lighthouse. The longest serving lighthouse in Australia, the current Macquarie opened in 1883, but the tunnel is purported to date only from the Second World War and its entrance has long since been covered up.

and Bondi Junction (half the original plan and some distance from the famous beach), but with three of its stations underground (Martin Place, Kings Cross and Bondi Junction). Two new platforms and short tunnel sections were also constructed at Sydney Central for a proposed Southern Suburbs Railway to Mascot, but this never came to fruition. The unused platforms were even allocated numbers 26 and 27 and sit quietly below the biggest station in Sydney despite never having seen a passenger. These 'ghost platforms' will now be used as communications rooms for the new Sydney Metro railway. There are half-built platforms beneath Redfern station – also intended for the Mascot Line.

The Airport Link built in the 1990s needed to be almost entirely underground, 4km (2.5 miles) was through rock and the remaining 6km (3.7 miles) through less difficult earth. Opened in the year 2000, it starts at Sydney Central, has two stations serving the airport and links with the Illawarra Line at Wolli Creek. In the northern suburbs, a new line was needed between Epping and Chatswood. The 12.5-km (7.7-mile) route is entirely in tunnel and opened in 2009, however it was temporarily closed just short of a decade later so that it could be integrated into the Sydney Metro Northwest project (see below).

Rapid transit

After decades of proposals – some of them fulfilled (City Circle) but many false starts – the new government decided, in 2011, to start Sydney's much needed rapid transit. As the city had piecemeal sections of subway all over the place, the Sydney Metro plan comes in several parts that will all be connected by 2024: Stage 1 is called Northwest; it uses the existing underground line between Epping and Chatswood, now extended at both ends. Tunnelling was completed in 2016 and there are now 15km (9 miles) of twin-bore tunnels lying up to 58m (190ft) deep from Epping to Bella Vista. The section from Bella Vista to Rouse Hill features an elevated section, called skytrain, from where it continues on the surface to the terminal at Tallawong. It opened in 2019. Further metro lines are being delivered to the Parramatta area and the new Western Sydney International (Nancy-Bird Walton) Airport.

ABOVE Museum station opened in 1926 with signage that echoed its London Underground counterpart (the large 'roundel' with station name on). At the time the plan was for a "City Circle" and several sub-surface lines, some of which were actually constructed but never used. When leaving Museum heading towards St. James it is just possible to make out the existence of some unused tunnelling.

RIGHT On the newly constructed Metro Northwest there is a huge chamber called Castle Hill Cavern. Now common in rail and road construction such crossovers are designed to enable vehicles to swap between each tunnel in case of emergency or for other operational reasons.

Stage 2 is the new tunnels from Chatswood that head south towards the city centre, for which boring began in 2018. Known as City and Southwest, it runs south through North Sydney, under the harbour and through the city centre towards Sydenham. It will require seven new underground stations before surfacing at Sydenham and running to Bankstown. This line provides eighteen stations in all on a 30km (18.6 mile) line with sections between 25–40m (82–130ft) deep.

Light-rail system

Sydney once had the largest tram system in the southern hemisphere, but it was all phased out by 1961. The city does now have the start of a modern light-rail system, however. The first line, L1, opened in 1997 and runs 12.8km (8 miles) with two underground stations. A new line, CBD and South East Light Rail runs to 12km (7.5 miles) long with nineteen stops (none underground, although there is a short tunnel under Moore Park). Several potential extensions are being studied, including to Parramatta, Bays Precinct, Green Square and the Anzac Parade corridor.

Road traffic

Developing infrastructure for car traffic has been just as challenging as establishing a rail network. The Cumberland County Plan proposed a radial motorway scheme as early as 1948, but little of it was in place by the early 1970s. Traffic patterns shifted to the south of the city and the airport, so the original plan was superseded by a 1987 proposal for which a number of projects were completed, some of them contentious.

Various extensions and widening schemes have led to the digging of many tunnels. The 900-m (2,950-yd) Sydney Harbour Tunnel runs 25m (82ft) below sea level and opened in 1992. The east–west Cross City Tunnel is 2.1km (1.3 miles) long and opened in 2005. The 3.6-km (2.2-mile) Lane Cove Tunnel, opened in 2007. A major scheme was the Eastern Distributor road tunnel, which was part of a 110-km (68-mile) Sydney Orbital Network. At 6km (3.7 miles) in length, it was built to link Sydney Airport to the Central Business District via the pre-existing Southern Cross Drive and opened in 1999. It needed 400,000m^3 (14,000,000ft^3) of spoil to be removed. Although much of it is in a trench, there is a 1.7-km (1-mile) 'piggyback' tunnel (one carriageway above another) just 32m (105ft) below one of Australia's most densely populated urban areas.

Another part of the Sydney Orbital Network is WestConnex, a 33-km (20-mile) predominately tunnelled road scheme providing a link between the M4 and M5 motorways. The last gap in the enormous orbital road will be filled in 2020 by NorthConnex, a 9-km (5.6-mile) tunnel linking the M1 and M2 motorways.

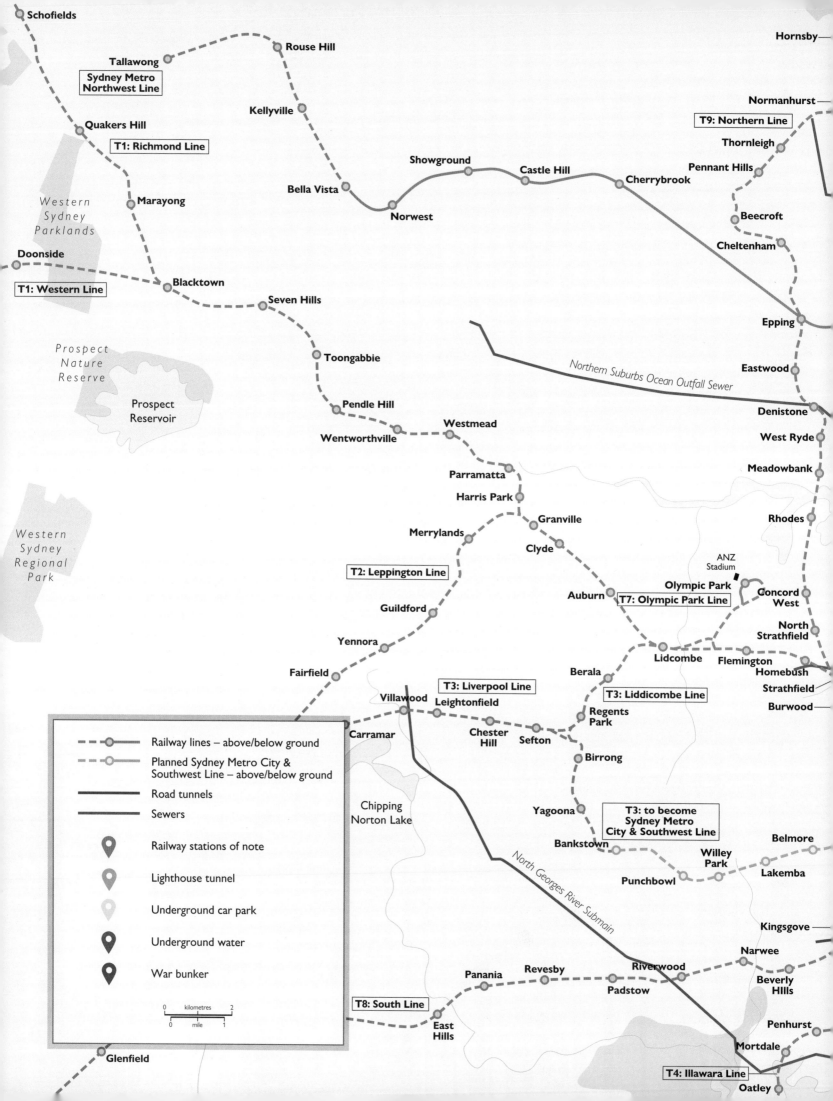

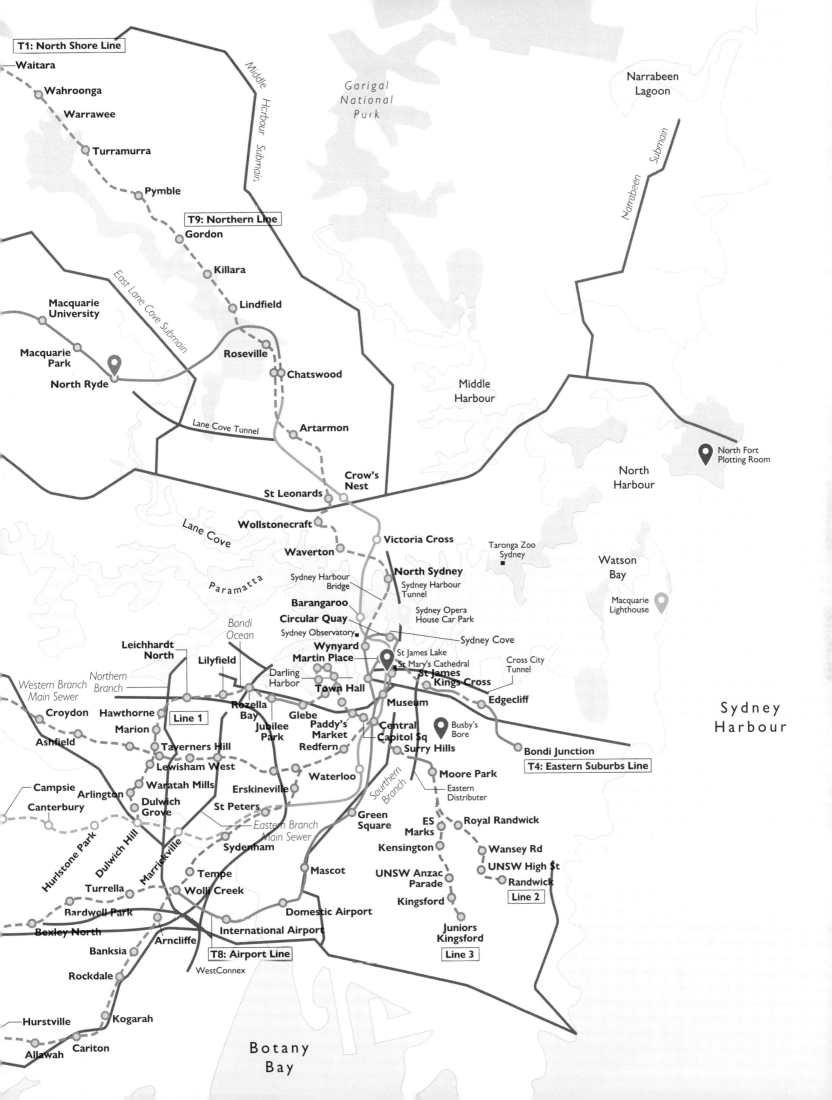

Honourable Mentions

The inclusion of the cities we have presented here is in no way a judgement on those that didn't make it into this tome. All books have space restrictions so the cities here are quite literally just 'scraping the surface': many other towns and cities with underground spaces exist, but there was simply just not enough room for all of them. It was a tough job editing down the entries to what was included, but honourable mentions must go to the following:

Albany, USA, which has the Empire State Plaza which features an underground police station.

Atlanta, USA, where an underground shopping and entertainment complex was opened in 1969.

Aydıntepe, Turkey, has a 3,000-year-old settlement carved out of volcanic rock.

Bakersfield, USA, features a series of tunnels which link the city to Tehachapi, 56km (35 miles) away.

Balaklava, Ukraine, which has a museum that was formerly an underground submarine base.

Birmingham, UK, has an underground telephone exchange, abandoned subterranean cinema, Royal Mail tunnel and numerous air raid shelters.

Boise, USA, where the state buildings are linked underground in the Capitol Mall Complex.

Bulla Regia, Turkey, an ancient ruin which features many houses that are partially underground.

Cleveland, USA, has a series of interconnected buildings called the Tower City Center.

Coober Pedy, South Australia, an intriguing town of dugout subterranean homes and hotels.

Corsham, UK, a 35-acre subterranean office and living space designated as Central Government War Headquarters.

Crystal City, USA, a district of Arlington in Virginia which is almost entirely subterranean.

Dallas, USA, home to a network of freifght tunnels as well as the Pedway connecting 36 blocks of shops, hotels and offices.

Derinkuyu, Turkey, once home to up to 20,000 people living up to 60m (197ft) underground (and connected by tunnels to Kaymakli).

Detroit, USA, was peppered with tunnels during the Prohibition.

Dover, UK, includes interconnected chalk tunnels mainly used for defensive purposes.

Duluth, USA, features many connected tunnels under the downtown area.

Edinburgh, UK, has many tunnels, as well as the Vaults, created in the empty voids of a bridge.

Frankfurt am main, Germany, has sub surface shopping malls named 'B-Ebene' and a tunnel for ships.

Fresno, USA, where 65 rooms of the Forestiere Underground Gardens provide relief from the heat at the surface.

Geneva, Switzerland, which is home to a large subterranean shopping centre linked to several buildings.

Glasgow, UK, where air raid shelters had to be built in the centre of roads owing to to the high density of population packed in to flats flanking them. One at Port Glasgow could house up to 1,000 people.

Guangzhou, China, has many sub-surface connecting tunnels between transport and offices.

Halifax, Canada, where tunnels link buildings in the downtown area.

Havre, USA, is a town that was rebuilt underground following a fire in 1904.

Hiroshima, Japan, where links between public transport are made below ground in the Kamiyachō Shareo.

Hole, Norway, the appropriately named town sits above the countries' largest underground war bunker and civilian shelter called the Sentralanlegget.

Houston, USA, where The Tunnel provides air-conditioned links between buildings above.

Indianapolis, USA, is home to many catacombs, which are used for cold storage.

Kandovan, Iran, where homes carved out of the cliffs are still inhabited.

Kansas City, USA is home to 'SubTropolis', the world's largest underground storage facility.

Kaymakli, Turkey, which boasts almost 100 tunnels and sheltering places, and linked to Derinkuyu.

Maastricht, Netherlands, has a vast linked cave system with over 20,000 corridors and a former gun emplacement that has been expanded with tunnels to act as a civil defence shelter.

HONOURABLE MENTIONS

Naours, France, in an old quarry are the tunnels once used as headquarters by the German occupation forces in the Second World War.

Naples, Italy, has many hidden catacombs, underground crypts and linking tunnels.

Nottingham, UK, sits on top of hundreds of man-made caves which date from the Middle Ages.

Oklahoma, USA, where the first tunnel of an underground pedestrian link between buildings was begun in the 1930s.

Osaka, Japan, has a handful of subterranean shopping malls – including one which houses over a thousand stores and restaurants.

Ottawa, Canada, has 5km (3 miles) of tunnels connecting shops and offices.

Ouyi, Iran, where the underground city is also called 'Noushabad' after its fresh water supply.

Petra, Jordan, probably the most famous city carved out of rock and dating back to 7000 BCE.

Philadelphia, USA, has miles of tunnels under the main streets linking to transit and important city offices.

Portland, USA, in which the Shanghai Tunnels were built to move goods up from the river into building basements.

Raleigh, USA, where The Village Subway, a former shopping and entertainment complex now lies abandoned.

Salt Lake City, USA has numerous tunnelled links between its churches.

San Francisco, USA has many abandoned freight and train tunnels.

Seattle, USA, where what are now sub surface passages were once at street level before the ground height was elevated.

Shenzen, China, has LinkCity which connects a number of underground malls to transit.

Singapore, the island country, is home to many underground shopping malls and connecting tunnels to transit stations, the largest of which CityLink has 5,575m² (60,000ft²) of retail space. It is also exploring the possibility of building an underground science park, called 'Science City'.

Seoul, South Korea, has two main underground shopping streets called Hoehyeon and Myeongdong and plans for many more.

Stockport, UK, four sets of underground air raid shelters dug though sandstone that could shelter up to 6,000 people.

Taipei, where there are underground streets and shops including CityMall beneath Civic Boulevard.

Tenjin, Japan, an underground shopping city in Fukuoka

Toledo, Spain, where caves were used by the ancients as baths, cemeteries and worship.

Washington, D.C., USA, has a large tunnelled system beneath the Capitol Complex and a miniature subway system for use of government employees only.

Winnipeg, Canada, where the Walkway is similar to Montreal and Toronto's underground malls.

Zurich, Switzerland, is home to RailCity: an underground shopping mall.

Further Resources

Bibliography

Ackroyd, Peter, *London Under, The Secret History Beneath The Streets*, Chatto & Windus, London, 2011

Alfredsson, Björn, et al, *Stockholm Under – 50 år-100 stationer*, Brombergs, Stockholm, 2000

Andreu, Marc, et al, *La ciutat transportada; Dos segles de transport collection al servei de Barcelona*. TMB, Barcelona, 1997

Bautista, Juan, et al, *Color Subterráneo*, Metrovias, Buenos Aires, 2007

Clairoux, Benoît, *Le Metro de Montreal*, Hurtubise HMH, Montreal, 2001

Collectiv, *Montréal en métro*, ULYSSE/STM, Montreal, 2007

Cudahy, B.J., *Under the sidewalks of New York; the story of the greatest subway system in the world*, Fordham University Press, New York, 1995

Dost, S. *Richard Brademann (1884–1965) Architekt der Berliner S-Bahn*, Verlag Bernd Neddermeyer, Berlin 2002.

Emmerson, Andrew and Beard, Tony, *London's Secret Tubes*, Capital Transport, Harrow Weald, 2004

Greenberg, Stanley, *Invisible New York: The Hidden Infrastructure of the City*, Johns Hopkins University Press, Baltimore, 1998

Hackelsberger, Christoph, *U-Bahn Architektur in München*, Prestel-Verlag, New York, 1997

Homberger, Eric, *The Historical Atlas of New York City: A Visual Celebration of 400 Years of New York City's History*, Henry Holt & Company, New York, 2005

Jackson, A. and Croome, D,. *Rails Through The Clay*, George Allen & Unwin, London, 1962

Lamming, Clive, *The Story of the Paris Metro*, Glenat, Paris, 2017

Macaulay, David, *Underground*, HMH Books for Young Readers, 1983

Marshall, Alex, *Beneath The Metropolis – The Secret Lives of Cities*, Carroll & Graff, New York, 2006

Moffat, Bruce G, *The "L" – The Development of Chicago's Rapid Transit System, 1888–1932*, Central Electric Railfans Assn, Chicago, 1995

New York Transit Museum, *The City Beneath Us: Building the New York Subway*, W. W. Norton & Company, 2004

Ovenden, Mark, *Paris Underground – The Maps, Stations And Design Of The Metro*, Penguin, New York, 2009

———, *London Underground by Design*, Penguin, London, 2013

———, *Transit Maps of the World*, Penguin Random House, London, 2015

———, *Metrolink – The First 25 years*, Rails Publishing/TfGM, 2017

Pepinster, Julian, *Le Metro de Paris – Plus d'un un Siecle d'Histoire*, La Vie du Rail, Paris, 2016

Price, Jane, *Underworld – Exploring The Secret World Beneath Your Feet*, Kids Can Press, Toronto, 2013

Schwandl, Robert, *Berlin U-Bahn Album*, Robert Schwandl Verlag, Berlin, 2002

———, *Metros in Spain*, Capital Transport, Harrow Weald, 2001

———, *München U-Bahn Album*, Robert Schwandl Verlag, Berlin, 2008

Strangest Books, *Strangest Underground Places in Britain*, Strangest Books, 2006

Talling, Paul, *London's Lost Rivers*, Random House Books, London, 2011

Warrender, Keith, *Below Manchester*, Willow Publishing, Altrincham, 2009

Webography

Please note: all the transit systems, most of the shopping malls and places that can be visited mentioned in this book have official operators websites which can easily be found with a simple online search. The list provided here is mainly of unofficial sites run by enthusiasts.

Canada

Montreal Metro fan: **emdx.org/rail/metro/index.php**
Montreal underground: **montrealundergroundcity.com**
Toronto PATH: **toronto.ca/explore-enjoy/visitor-services/path-torontos-downtown-pedestrian-walkway/**
Toronto Subway fan: **transit.toronto.on.ca**

China

Rail & Metro fans: **en.trackingchina.com**

Finland

Helsinki underground masterplan: **hel.fi/helsinki/en/housing/planning/current**

France
French subterranean spaces: **souterrains.vestiges.free.fr**
Paris catacombes explorer: **annales.org/archives/x/gillesthomas.html**
Paris Metro curiosities: **paris-unplugged.fr/category/metro-2/**
Paris Metro fan site: **siteperso.metro.pagesperso-orange.fr**
Paris Metro preservation society: **ademas.assoc.free.fr**

Germany
Berlin underground fan: **berliner-unterwelten.de/en.html**
Munich U-Bahn fan: **u-bahn-muenchen.de**
U-Bahn Archive: **u-bahn-archiv.de/**
U-Bahn fan: **berliner-u-bahn.info**

Hungary
Metro fan site: **metros.hu**

India
Transit fan: **themetrorailguy.com**
Urban transit news: **urbantransportnews.com**

Italy
Milan Metro fan: **sottomilano.it**
Rail fans: **cityrailways.com**
Rome underground fan: **sotterraneidiroma.it/en**

Mexico
Metro fan: **mexicometro.org**

Russia
Metros across Russia fan site: **meta.metro.ru**
Moscow Metro fan site: **metro.ru**
Other Russian Metros fan site: **mirmetro.net**

South America
Buenos Aires Subte fan: **enelsubte.com**
Latin America Subways: **alamys.org/es/**

Spain
Metro fans: **anden1.es**
Train fans: **trenscat.com**

UK
Abandoned stations: **disused-stations.org.uk**
Closed Mail Rail: **postalmuseum.org/discover/attractions/mail-rail/**
Hidden London tour: **ltmuseum.co.uk/whats-on/hidden-london**
Hidden Manchester tour: **hidden-manchester.org.uk/tunnels.html**
Liverpool tunnels: **williamsontunnels.co.uk/view.php?page=about**
Subterranea Britannica: **subbrit.org.uk**
Underground fan: **londonreconnections.com**
Underground forum: **districtdavesforum.co.uk**

USA
Chicago El: **chicago-l.org**
Chicago Pedway: **chicagodetours.com/images/chicago-pedway-map-detours.pdf**
LA Transit Coalition: **thetransitcoalition.us/RedLine.htm**
NYC Subway: **nycsubway.org**
NYC Transit: **rapidtransit.net**
Transit advocates news: **thetransportpolitic.com**

Worldwide
All urban rail systems: **urbanrail.net**
Forbidden places: **forbidden-places.net**
Metro data: **mic-ro.com/metro/**
Strange places: **atlasobscura.com**
Subways forum: **skyscrapercity.com/forums/subways-and-urban-transport.130/**
Transit maps: **transitmaps.tumblr.com**
Underground explorers: **undergroundexplorers.com**

Wikipedia
en.wikipedia.org/wiki/List_of_metro_systems
en.wikipedia.org/wiki/Underground_city
en.wikipedia.org/wiki/Urban_exploration

Index

A
AEG experimental rail tunnel, Berlin 136, 152–153
AHR consultancy practice 99
Albert Cuyp Car Park, Amsterdam 128–129
Alfonso XII of Spain 82
American revolutionary War 64
Amsterdam, Netherlands 126–131
    Albert Cuyp Car Park 128–129
    bomb shelters 51
    cellars 51
    IJ tunnel 126
    map of sub-surface systems 130–131
    rapid transit 126–127
    waterways 126
    Weesperplein Station 127, 128
Apsītis, Vaidelotis 186
Aqua Virgo, Rome 147
Aqua-Line Tunnel, Tokyo Bay 200, 204
archaeology
    Amsterdam 126
    Barcelona 104, 105
    Beijing 196, 199
    Buenos Aires 72
    Gibraltar 80
    London 94–95
    Madrid 82
    Mexico City 21, 22
    Milan 134
    Moscow 182
    Mumbai 194–195
    Paris 112
    Rome 142–147
    Rotterdam 125
    Sweden 168–169
    Sydney 208
    Tokyo 201
Arquitectoma 21
Augustus 142
Autopista de Circunvalacíon M-30, Madrid 85
Ayotte et Bergeron 51
Aztecs 18, 21, 22

B
Balázs, Mór 164
Barbosa, Alfons Soldevila 107
Barcelona, Spain 104–111
    bomb shelters 108
    Llefia Station 106–107, 108
    map of sub-surface systems 110–111
    Metro 105–108
    Diamond Square Air-raid Shelter 106–107
    Roman roots 104
    urban planning 104–105
    water collection tanks 108
Bazalgette, Joseph 98, 99

Beach, Andrew Ely 55, 57
Beijing, China 196–199
    ancient treasures 199
    Dìxià Chéng 197
    sanitation and water needs 199
    subway plans 198–199
Belgrand, Eugène 113
Berlier, Jean-Baptiste 114
Berlin, Germany 100, 152–159
    Berlin Wall 156
    combined and separate sewer systems 154–155
    Hitler's legacy 153–156
    Lindentunnel 152
    map of sub-surface systems 158–159
    reunification 156–157
    railway tunnels 152–153
    water management 152
Besteiro, Julián 83
Biancoshock 136, 137
Bibliothèque National de France, Paris 119
Bienvenue, Fulgence 115
bike vaults, Tokyo 202–203, 204
Bonaparte, Napoleon 112, 113, 183
Bonaparte, Joseph 82
Boring Company 13, 14–15
Borioli, Baldassare 135
Boston, USA 31, 54, 64–71
    Big Dig 68
    early rail and road networks 64–68
    map of sub-surface systems 70–71
    old streams 64
    original brick sewer network 64
    Steinert Hall 64, 66
    Ted Williams Tunnel 66–67, 68
Bradfield, John 209
Brame, Édouard 114
Brand, Bill Masstransiscope 60
breweries 61, 127, 151, 165
British East India Company 194
Broggi, Carlo 136
Brunel, Marc Isambard 101
Bucharest, Romania 137
Budapest, Hungary 100, 160–167
    Hospital in the Rock 163
    Kálvin tér Station 162–163
    karst landscape 160–164
    Kőbánya Mine 165
    map of sub-surface systems 166–167
    Metró development 164–165
    Rákosi-bunker 165
Buenos Aires, Argentina 72–77
    El Zanjon de Granados 72, 74–75
    Manzana de las Luces tunnels 72
    map of sub-surface systems 76–77

    Paseo del Bajo Road Corridor Project 74–75
    Subte 72–73
Bunker 42, Moscow 186–187, 189
Bunker Arquitectura 23
Bunker Hill, Los Angeles 11, 12–13
Burnham, Andy 91
Bury, Thomas Talbot 87
Busby, John 208
Busby's Bore, Sydney 208, 209
buses 12, 133
    Los Angeles 10–11

C
Calumet TARP Pumping Station, Chicago 33
canals 90
    Amsterdam 126
    Canal de Marseille au Rhône 132
    Rotterdam 122–125
Candiani, Leopolodo 135
Cané, Miguel 73
Carey, Hugh L. 55
cars 12
    Albert Cuyp Car Park, Amsterdam 128–129
    underground car parks, Tokyo 201–204
    Vatican City Car Park, Rome 147
Casa Loma, Toronto 43
Castiglioni, Carlo 135
catacombs 61, 112, 147
Cathedral Steps, Manchester 92, 93
Cerdà, Ildefons 104–105
Champlain, Samuel de 46
Châtelet-les-Halles, Paris 116–117
Chesbrough, Ellis S. 26
Chester, UK 86
Chicago, USA 26–35, 38, 58
    cable cars 27–30
    Calumet TARP Pumping Station, Chicago 33
    Clark/Lake Station 28–29
    elevating the city 26
    freight train tunnels 28–29, 31–32
    Great Chicago Fire 1871 27
    map of sub-surface systems 34–35
    Pedway 13, 32–33
    rail innovations 30–32
    tunneling beneath the river 26–27
Christ 142
Churchill War Rooms, London 100, 101
Cincinnati, USA 36
    Hopple Street Tunnel 37
    thwarted subway plans 36–37
Clark/Lake Station, Chicago 28–29
Cloaca Maxima, Rome 142
Close, Chuck 56
Cold War 12, 81, 92, 125, 127, 156, 160, 164, 165, 176, 182, 186, 197

INDEX

Colosseum, Rome 142, 144–145, 146
Connell, William 93
Cook, James 208
Cook, Michael 42
Cortés, Hernán 18
Crillon, Duc de 81
Crossrail, London 100
Crown Finish Caves, New York City 61
crypts 61, 134

D
Diamond, Robert 55
Diamond Square Air-raid Shelter 106–107
disease 26, 80, 95–98, 112
Dixià Chéng, Beijing 197
Domus Aurea, Rome 145, 146
Donskoy, Dmitri 182
Dutch East India Company 122, 126
Dutch West India Company 54

E
earthquakes 22–23, 200, 205
earthscrapers 22–23
Eckstein, Jorge 72
Edison 134, 136
Eiffel Tower 112, 115
Eiffel, Gustave 115
Engel, Ludvig 176
Erasmusbrug, Rotterdam 123–125
Expo 67, Montreal 47, 50
Exposition Universelle 1889, Paris 115

F
First World War 55, 113, 136, 183
Flachat, Eugène 114
flooding 18, 32
    Barcelona 108
    Great Flood 1913, Ohio 36
    Munich 151
    Paris 113
    Rotterdam 122
    Tokyo 200–201
food storage 165
Forum des Halles, Paris 116–117
Foshan Xinjing Industrial Company 198
Franco, Francisco 82

G
Gaudí, Antoni 105
Gehry, Frank 10
Gellért Hill, Budapest 160, 164
ghost stations 60, 73, 105–108, 115, 125, 139–140, 153, 156–157, 173, 204–205
Gibraltar 80
    clean water 80–81
    Great Siege Tunnels 80, 81
    strategic base 81
Glasgow, UK 100
gold reserves 40, 43, 60
Granche, Pierre Système 50
Grand Paris Express 118–119
Grease 10
Great Depression 36, 50, 60
Grimshaw Architects 39
Guardian Underground Telephone Exchange (GUTE), Manchester 92–93
Guinness World Records 55
Guzmán, Juan Alonso de 80

H
Haakon V of Norway 138
Hallek, Enno 172
Hancox, Joy 93
Hanseatic League 169
Harald Hardrada 136
Haussmann, Georges 113
Heins & Lafarge 57
Helsinki, Finland 176–181
    Cold War anxiety 176
    early history 176
    Helsinki Underground Master Plan 177
    Itäkeskus Swimming Hall 178–179
    Kamppi Station 178–179
    map of sub-surface systems 180–181
    metro system 176–177
Henriksdal Wastewater Treatment Plant, Stockholm 169–171
Hitler, Adolf 153–156
Hobrecht, James 152
Holden, Charles 97
Hollywood, USA 10
Hollywood/Highland Station, Los Angeles 13
Hollywood Vine Station, Los Angeles 14–15
Hospital in the Rock, Budapest 160–164
Hunt, Henry Thomas 36
Huntingdon, Henry 11–12
Hurricane Sandy (2012) 55

I
ice storage 61, 95
Ingalls Building, Cincinnati 36
Iron Curtain 156, 176
Itäkeskus Swimming Hall, Helsinki 178–179
Ivan the Terrible of Russia 182

J
John 86
John Mowelm 97
Julius Caesar 142

K
Kálvin tér Station, Budapest 162–163
Kamppi Station, Helsinki 178–179
Klein, Sheila 13
KMD Arquitectos 21
Kremlin, Moscow 182–183

L
Labelle, Marchand et Geoffroy 50
Leonardo da Vinci 134
LFO Arkitektur og Design 139
Liniers, Santiago de 72
Liverpool, UK 31, 54, 55, 86–89
    developing railway network 87–88
    Williamson's Tunnels 88–89
    world's first major underground rail infrastructure 86–87
Llefia Station, Barcelona 106–107, 108
London, UK 38, 58, 94–103, 113–114
    London Silver Vaults 101
    London Underground 98–100
    map of sub-surface systems 102–103
    Piccadilly Circus Station 96–97
    Roman roots 94–95
    super-sewer 95–98
    unique feats of engineering 100–101
Longueil, Canada 50
Los Angeles, USA 10–17
    automobile traffic 12
    Boring Company test tunnel 14–15
    Bunker Hill 11, 12–13
    Hollywood/Highland Station 13
    Hollywood/Vine Station 14–15
    Hyperloop 13
    Los Angeles Aqueduct 10, 11
    map of sub-surface systems 16–17
    Metro 12
    Regional Connector 13
    streetcars 10–11
    subways 11–12
    water management 10
Louvre, Paris 112, 119
Ludwig I of Germany 150
Luján, Gilbert (Magu) 15

M
Maastunnel, Rotterdam 122, 123
Madrid, Spain 82–85
    city transport 83–85
    hill tunnels 82
    Royal Palace tunnels 82
    Spanish Civil War 82–83
Maeslantkering, Rotterdam 124, 125
Manchester, UK 54, 58, 90–93
    canals 90
    hidden gems 92–93
    Manchester & Salford Junction Canal 90, 91
    scuppered metro plans 90–91
Manzana de las Luces, Buenos Aires 72, 76
Mao Zedong 197
maps of sub-surface systems
    Amsterdam 130–131
    Barcelona 110–111
    Berlin 158–159
    Boston 64–65, 70–71
    Budapest 166–167
    Buenos Aires 76–77
    Chicago 34–35
    Helsinki 180–181
    London 102–103
    Los Angeles 16–17
    Mexico City 24–25
    Montreal 52–53
    Moscow 190–191
    New York City 62–63
    Paris 120–121
    Rome 148–149
    Stockholm 174–175
    Sydney 214–215
    Tokyo 206–207
    Toronto 44–45
Marseille, France 132–133
    Canal de Marseille au Rhône 132
    road tunnels 133
    towards a metro system 132–133
Mass Transit Railway, Hong Kong 198
McQuarrie Lighthouse, Sydney 211
Meiji of Japan 200
Merola, Mario 51
Mexico City, Mexico 18–25
    early transportation 18–19
    Garden Santa Fe 20–21, 22
    mammoth bones at Talisman Station 20–21, 22
    map of sub-surface systems 24–25
    Metro 19–22
    sewage solutions 18
    structural wonders 22–23
Miami and Erie Canal, USA 36
Milan, Italy 134–137
    Biancoshock 136, 137
    historic remains 134
    Metropolitana 135–137
    Piazza Oberdan 136, 137
    trams 134–135

Minorini, Franco 136
Minuit, Peter 54
Mira, Carlo 135
Miró, Joan 108
Molnár János caves, Budapest 160
Montreal, Canada 40, 46–53
    early city development 46
    map of sub-surface systems 52–53
    Metro 50–51
    RÉSO 48–49
    subterranean waterworks 46–47
    Underground City 47–50
Monza, Italy 134, 136
Moscow, Russia 182–191
    bunkers 42 186–187, 189
    Kremlin 182–183
    map of sub-surface systems 190–191
    Metro 183–189, 198
    Rzhevskaya Station 186–187
Moses, Robert 56
Mumbai, India 194–195
    ancient tunnels 194–195
    reclamation 194
    transportation 195
Munich, Germany 150–151
    U-Bahn 150–151
    water management 151
Municipal Ossuary, Paris 116–117
mushroom growing 165
Musk, Elon 13
Mussolini, Benito 147

N
Napoleon III of France 113
Nationaltheatret Station, Oslo 139, 140
Neanderthals 80
necropolises 146–147
Nero 143, 145, 146
New York City, USA 31, 54–63, 64, 100
    6½ Avenue 60
    cheese storage 61
    city history 54
    early railroad tunnels 55–56
    ghost stations 60
    ice storage 61
    map of sub-surface systems 62–63
    mass public transport 56–57
    McCarren Park pool complex 60
    New York Steam Company 54–55
    Second Avenue Line 57–60
    St Patricks Old Cathedral, Soho 61
    Times Square–42nd Street Station 58–59
    Wall Street cylindrical vault 60
    West Side Cow Pass 61
New York Steam Company 54–55
Nieuwe Maas River, Rotterdam 122–125
nuclear shelters 12, 37, 47, 125, 127, 156, 160, 164, 165

O
Olympic Games 10, 47, 151, 176, 198
Oslo, Norway 138–141
    Opera Tunnel system 140–141
    rail network 138–140
ossuaries 116–117, 147
Osterrath, Pierre 51

P
Pacific Electric Railway, Los Angeles 11–12
Palacios, Antonio 84

Palais Garnier underground lake, Paris 119
Pallarp, Åke 172
Pape, Chris 'Freedom' 56
Paris, France 38, 46, 58, 100, 112–121, 147
    Bibliothèque Nationale de France 119
    floods and clean water 113
    Forum des Halles 116–117
    Louvre 112, 119
    map of sub-surface systems 120–121
    Métro 31, 50, 113–115
    Municipal Ossuary 116–117
    quarries 112
    Second World War 115, 119
    serving the suburbs 118–119
    underground lake 119
Parsons, William Barclay 57
Paseo del Bajo Road Corridor Project, Buenos Aires 74–75
PATH, Toronto 40–41, 42–43
Paul Raff Studio 39
Pearson, Charles 99
Pedway (Downtown Pedestrian Walkway System), Chicago 13, 32–33
Pellatt, Sir Henry 43
Peter the Great of Russia 182
Philadelphia, USA 31, 54, 58
Piccadilly Circus Station, London 96–97
pneumatic mail systems 38, 55–56, 191
Ponte, Vincent 47
Privately Owned Public Spaces (POPS), New York City 60–61
psychiatric hospitals 43

Q
Queensway Tunnel, Liverpool 88, 89
Quintana, Bernardo 19

R
railways
    Beijing 199
    Berlin 152–153
    Chicago's Elevated (L) 30–32
    Liverpool 86–88
    London 98–99, 101
    Madrid 82, 85
    Mexico City 18–19
    Mumbai 195
    Munich 150
    New York City 55–56
    Oslo 140
    Paris 113–115, 118–119, 165
    Rome 147
    Sydney 208–212, 213
    Tokyo 204
Raj Bhavan bunker, Mumbai 195
Rákosi, Mátyás 165
Rat Tribe, Beijing 197
Reagan, Ronald 13
Reinfelds, A. 186
Réseau Électrique Métropolitain (REM), Montreal 50–51
Réseau Express Régional (RER), Paris 118–119, 165
RÉSO, Montreal 47–50
roads
    Central Artery/Tunnel Project (CA/T), Boston 68
    Gibraltar 81
    Madrid 85
    Marseille 133
    Oslo 140–141
    Paseo del Bajo Road Corridor Project, Buenos Aires 74–75

    Rotterdam 123–125
    Stockholm 173
    Sydney 213
    Ted Williams Tunnel, Boston 66–67, 68
    Tokyo 200
Rome, Italy 142–149
    ancient history 142
    car parks 147
    cities of the dead 146–147
    Cloaca Maxima 142
    Colosseum 142, 144–145, 146
    Domus Aurea 145, 146
    Great Fire (64 CE) 143
    map of sub-surface systems 148–149
    Metropolitana 147
    underground ruins 143
Roosevelt, F. D. 60
Rotes Rathaus, Berlin 157
Rotterdam, Netherlands 122–125
    Nieuwe Maas River 122–125
    storm protection 125
    subterranean remains 125
    Time Stairs 125
Rzhevskaya Station, Moscow 186–187

S
Saarinen, Eliel 176
Saint Basil's Cathedral, Moscow 182–183
Second Avenue Line, New York City 57–60
Second World War 12, 46, 60, 81, 90, 92, 100, 101, 115, 119, 122, 123, 134, 135, 147, 150, 153–156, 173, 182
Seven Years' War 38
sewers
    Beijing 199
    Berlin 152, 154–155
    Boston 64
    Buenos Aires 72
    Chicago 26
    London 95–98
    Mexico City 18
    Montreal 46–47
    Paris 113
    Rome 142
    Stockholm 169–171
    Tokyo 201
    Toronto 42
shopping malls
    Garden Santa Fe, Mexico City 20–22
    Toronto 40–41, 42–43
Skeppsholmen Caverns, Stockholm 173
Sky Mile Tower, Tokyo 205
slave trade 64, 86
Spanish Civil War 81, 82, 108
Speer, Albert 156
St Petersburg, Russia 176, 182
St. Peter 146
St. Stephen 160
Stalin, Josef 183, 188
Statue of Liberty, New York City 54
Stefini, Evaristo 136
Steinert Hall, Boston 64, 66
Stephenson, George 86, 87
Stockholm, Sweden 168–175
    city transportation 169–173
    Henriksdal Wastewater Treatment Plant 170–171
    map of sub-surface systems 174–175
    mountain room (Bergrummet) 173
streetcars
    Chicago 27–30

Cincinnati 36
Los Angeles 11–12
Mexico City 18–19
Toronto 38
Suárez, Esteban 23
subways
    Amsterdam 126–129
    Barcelona Metro 105–108
    Beijing Subway 198–199
    Berlin U-Bahn 153, 156–157
    Boston 64–68
    Budapest Metró 164–165
    Buenos Aires Subte 72–73
    Chicago 32
    Cincinnati 36–37
    Helsinki Metro 176–179
    London Underground 98–100
    Los Angeles Metro 12
    Madrid Metro 82, 83–85
    Manchester 90–91
    Marseille 132–133
    Metropolitana di Milano 135–137
    Metropolitana di Roma 147
    Mexico City Metro 19–22
    Montreal Métro 50–51
    Moscow Metro 183–189, 198
    Mumbai Metro 195
    Munich U-Bahn 150–151
    New York City Subway 56–60
    Oslo Tunnelbanen 139–140
    Paris Métro 31, 50, 113–115
    Réseau Électrique Métropolitain (REM), Montreal 50–51
    Rotterdamse Metro 122–123
    Stockholm Tunnelbana 169, 172–173
    Sydney Metro 212–213
    Tokyo Metro 204–205
    Toronto Subway 38–42
Svensson, David 173
Sydney, Australia 208–215
    early infrastructure 208
    light-rail system 213
    Macquarie Lighthouse 211
    map of sub-surface systems 214–215
    rapid transit 212–213
    reaching the suburbs 209–212
    road traffic 213
    Sydney Harbour Bridge 208–209
    Sydney Opera House Car Park 210–211

T
Tallin, Estonia 176
Tarquinius Superbus 142
Ted Williams Tunnel, Boston 66–67, 68
Thames Tideway Tunnel, London 98
Thames Tunnel, London 100–101
Thury, Louis-Étienne Héricart de 117
Tijdtrap, Rotterdam 125
Times Square–42nd Street Station, New York City 58–59
Tokyo, Japan 23, 200–207
    bike vaults 202–203, 204
    earthquakes and flooding 200–203
    hidden city 205
    map of sub-surface systems 206–207
    Tokyo Skytree 205
    urban rail network 204–205
    world's deepest foundations 205
Toronto, Canada 38–45, 50
    early infrastructure 38
    Casa Loma tunnels 43
    Great Fire (1904) 38
    Lakeshore psychatric hospital tunnels 43
    map of sub-surface systems 44–45
    PATH 40–41, 42–43
    subterranean shopping 42, 43
    subway 38–42
    underground waterways 42, 43
    Vault, The 40
Tower Subway, London 101
trams 12
    Barcelona 105
    Berlin 152
    Boston 64–65
    Budapest 164
    Buenos Aires 73
    Madrid 83
    Marseille 132
    Milan 134–135
    Mumbai 195
    Munich 150
    Oslo 138
    Rotterdam 122, 125
    Stockholm 169–172
    Sydney 213
tsunamis 200
Tunel Emisor Oriente (TEO), Mexico City 18
Tunnel du Rove, Marseille 132

U
Ultvedt, Per-Olof 172
Underground City, Montreal 47–50
USSR (Union of Soviet Socialist Republics) 182

V
Vatican City, Rome 142, 146, 147
Vaughan Metropolitan Centre, Toronto 39
Vault, The, Toronto 40
vegetable growing 101
Vezér, Ferenc 160

W
Warhol, Andy 60
water management 108, 126, 151, 200–201
water supply
    Beijing 199
    Chicago 26
    Gibraltar 80–81
    Los Angeles 10, 11
    Montreal 46–47
    Toronto 42, 43
Weesperplein Station, Rotterdam 127, 128
Who Framed Roger Rabbit 12
Willem IV of Holland 122
Williams, Arthur Robert Owens 81
Williamson, Joseph 88–89
Woolwich foot tunnel, London 101
Workhorse and PAC 60
Wren, Christopher 95
Wright, Frank Lloyd 169
Wyman, Lance 22

Y
Yamate Tunnel, Tokyo 173, 200

Z
Zanjon de Granados, Buenos Aires 72, 74–75
Zero Carbon Food 101

Credits

Acknowledgements

Author's appreciation for their assistance on compiling this book: Richard Archambault, Mike Ashworth, Jennifer Barr, Laura Bulbeck, Luca Carenzo, Pat Chessell, Roman Hackelsberger, Leo Frachet, Kate Gunning, Reka Komoli, Juan Loredo, Peter B. Lloyd, Geoff Marshall, Julian Pepinster, Maxwell Roberts, Chris Saynor, Rob Shepherd, Julia Shone, Guy Slatcher, Anna Southgate, Paul Talling, Adam Wales, Mike Walton, Lucy Warburton.

Picture Credits

Map Illustrations by Lovell Johns.

Underground Feature Illustrations © 2020 Robert Brandt.

The illustrations are intended to be artistic interpretations, but cannot claim to be exact representations. Some were adapted from reference material from these sources: **front cover & 116–17** Patrick Berger and Jacques Anziutti Architectes; **20–21 b** KMD Architects; **28 l & 58–59** Candy Chan (Project Subway NYC); **96–97** TFL/public domain; **106** Soldevila Arquitectes; **123** *Maasbode*, 1938; **128 b** Architectur Centrum Amsterdam/Architects: B. Spängberg & S. van Rhijn; **154–55** Berliner Wasserbetriebe; **162–63 b** Palatium Studio; **170–71** Mario Salutskij, Illustrerad Teknik; **178–79** Hkp Architects; **186–87 b** Blank Architects.

All other images:

9 Chronicle/Alamy Stock Photo; **11 l** Courtesy of the Library of Congress, HAER CA-298-AH; **11 r** Dorothy Peyton Gray Transportation Library and Archive at the Los Angeles County Metropolitan Transportation Authority; **13** Courtesy of the Library of Congress, LC-DIG-highsm-24291; **19** Susana Gonzalez/Bloomberg via Getty Images; **22** ALFREDO ESTRELLA/AFP via Getty Images; **23** Gengiskanhg/CC BY-SA 3.0; **27 t** Everett Collection Historical/Alamy Stock Photo; **27 b** Chicago Sun-Times/Chicago Daily News collection/Chicago History Museum/Getty Images; **30** Chicago Sun-Times/Chicago Daily News collection/Chicago History Museum/Getty Images; **31** Pictures Now/Alamy Stock Photo; **32** Heritage Image Partnership Ltd/Alamy Stock Photo; **33 t** stevegeer/Getty Images; **33 b** UrbanImages/Alamy Stock Photo; **39** Roberto Machado Noa/LightRocket via Getty Images; **42** Bloomberg/Getty Images; **43** Lissandra Melo/Shutterstock; **47** The Cosmonaut/CC BY-SA 2.5 CA; **50** art_inthecity/CC BY 2.0; **51** art_inthecity/CC BY 2.0; **56** Metropolitan Transportation Authority/Patrick Cashin/CC BY 2.0; **57** Michael Freeman/Alamy Stock Photo; **61** Randy Duchaine/Alamy Stock Photo; **65 t** Courtesy of the Library of Congress, LC-D4-33302; **65 b** George H. Walker & Co.; **68** Xb-70/public domain; **69** David L Ryan/The Boston Globe via Getty Images; **69** Hellogreenway/CC BY-SA 3.0; **73** Luis Marden/National Geographic Image Collection/Bridgeman Images; **79** Universal History Archive/Universal Images Group via Getty Image; **84** agefotostock/Alamy Stock Photo; **85** agefotostock/Alamy Stock Photo; **87** The Stapleton Collection/Bridgeman Images; **88 t** © Illustrated London News Ltd/Mary Evans; **88 b** Trinity Mirror/Mirrorpix/Alamy Stock Photo; **91** Popperfoto via Getty Images; **92** Trinity Mirror/Mirrorpix/Alamy Stock Photo; **95** Frederick Wood - Punchy/Alamy Stock Photo; **98** SSPL/Getty Images; **99** Copyright Crossrail Ltd; **100** Alex Segre/Alamy Stock Photo; **105** Endless Travel/Alamy Stock Photo; **108** Javierito92/CC BY 3.0; **109 t** Biblioteca MUHBA/CC BY 2.0; **109 b** TMB Archive; **113** PATRICK KOVARIK/AFP/GettyImages; **114 t** Eugène Trutat/Muséum de Toulouse (public domain); **114 b** Mary Evans/Grenville Collins Postcard Collection; **118** Perry van Munster/Alamy Stock Photo; **119** agefotostock/Alamy Stock Photo; **124 t** Henk Lindeboom/Anefo/Dutch National Archives/CC0 1.0; **124 b** A G Baxter/Shutterstock; **125** frans lemmens/Alamy Stock Photo; **127** David Peperkamp/Shutterstock; **133** laranik/Shutterstock; **136** Franco Ricci/Alamy Stock Photo; **137 t** Pandarius/Alamy Stock Photo; **137 b** © Biancoshock; **139** Lukas Bischoff Photograph/Shutterstock; **141 t** Punkmorten/public domain; **143** Werner Forman/Universal Images Group/Getty Images; **146 t** Sotterranei di Roma; **146 b** Universal History Archive/Universal Images Group via Getty Images; **151 b** lukas bischoff/Alamy Stock Photo; **153 tl** © Berliner Unterwelten e.v./H. Happel; **153 tr** Purschke/ullstein bild via Getty Images; **153 bl** © Berliner Unterwelten e.v./H. Happel; **153 br** © Berliner Unterwelten e.v./H. Happel; **156** © Berliner Unterwelten e.v./H. Happel; **157** Collignon Architektur; **161 t** Nature Picture Library/Alamy Stock Photo; **161 b** Acceptphoto/Alamy Stock Photo; **164** Jakub Dvořák/Alamy Stock Photo; **165** Hercules Milas/Alamy Stock Photo; **168–69** Julian Herzog/CC BY 4.0; **173** Manuel Bischof/Getty Images; **177 t** City of Helsinki Media Bank; **177 b** Carl Mydans/The LIFE Picture Collection via Getty Images; **183** Vladimir Mulder/Shutterstock; **184 t** A.Savin/ Free Art License; **184 b** Florstein/CC BY-SA 4.0; **185 t** Antares 610/CC0 1.0; **185 b** Vermette108/Shutterstock; **188** Eddie Gerald/Alamy Stock Photo; **189** Lagutkin Alexey.Shutterstock; **193** Historic Collection/Alamy Stock Photo; **194** British Library/public domain; **196–97** USGS/public domain; **198** Shan_shan/Shutterstock; **199** 颐园新居/CC BY-SA 3.0; **204** Daboost/Shutterstock; **205** Kyodo News Stills via Getty Images; **209** George Lipman/Fairfax Media via Getty Images; **212** MMXeon/Shutterstock; **213** Cath Bowen/Rusty Goat Media for Sydney Metro.